Precision Crystallization

Theory and Practice of Controlling Crystal Size

T0179295

Precision Crystallization

Theory and Practice of Controlling Crystal Size

Ingo H. Leubner, Ph.D.

CRC Press
Taylor & Francis Group
Boca Raton   London   New York

CRC Press is an imprint of the
Taylor & Francis Group, an **informa** business

CRC Press
Taylor & Francis Group
6000 Broken Sound Parkway NW, Suite 300
Boca Raton, FL 33487-2742

First issued in paperback 2019

ISBN-13: 978-1-4398-0674-6 (hbk)
ISBN-13: 978-0-367-38516-3 (pbk)

Library of Congress Cataloging-in-Publication Data

Leubner, Ingo H.
    Precision crystallization : theory and practice of controlling crystal size / Ingo H. Leubner.
        p. cm.
    Includes bibliographical references and index.
    ISBN 978-1-4398-0674-6 (alk. paper)
    1. Crystal growth. 2. Nucleation. 3. Crystallization. I. Title.

QD921.L48 2010
660'.284298--dc22

2009020571

Visit the Taylor & Francis Web site at
http://www.taylorandfrancis.com

and the CRC Press Web site at
http://www.crcpress.com

Contents

Preface

Crystalline materials make up an estimated 80% of chemical and pharmaceutical products, and their crystal size and size distribution are often important factors for product applications. Yet, no practical guidance has been available to the chemist and product engineer to achieve the precision control of the size and size distribution of their crystalline materials using practical and available control variables.

During 30 years in the precision precipitation of crystals for product applications, the author supervised and executed more than 1,500 controlled precipitations. In this work, he found that the classical models did not provide practical information to guide him in his experiments. Thus, he and his colleagues, Jong Wey and Ramesh Jagannathan, developed and published a novel science-based practical model to control crystal nucleation, and thus crystal size and size distribution. This model distinguishes itself from previous nucleation models by including both nucleation and growth in the nucleation process, thus the name *Balanced Nucleation and Growth* (BNG) model. He continued the development of the model and provided experimental support. The new model, which is the core of this book, gives parameter-free predictions for both controlled batch and continuous crystallizations.

For the first time, the results of the nucleation process are quantitatively related to practical experimental control variables such as reactant addition rate, crystal solubility, temperature, residence time (continuous crystallizations), and the effect of ripening agents (crystal-supersizing) and crystal growth restrainers (crystal nanosizing) during nucleation. In addition, the theory predicted previously unknown phenomena and corrected erroneous perceptions on the importance of reaction volume on the outcome of crystal nucleation. The model, its correlations with control parameters, and other predictions were supported by experimental results. This sets this model apart from many other modeling efforts.

This book offers the concepts and quantitative models that are vital to those who need precision control of crystal size in products. The crystal size is a function of both crystal nucleation and growth, where control of nucleation is the most challenging factor. Classical nucleation theories generally do not give precise guidance to control crystal nucleation and solutions to specific problems are generally obtained by trial and error.

The models and equations in this book relate the crystal number and size distribution (nucleation) to experimentally controlled reaction variables. Crystal number and size are quantitatively related to reactant addition rate, crystal solubility, temperature, and solvent and crystal properties. It also models the effect of other factors such as crystal ripening agents and crystal growth restrainers for the first time; equations for both controlled batch and continuous precipitations are developed using the same model. Unexpected predictions were experimentally confirmed.

The new concepts have been applied to the precipitation of inorganic materials, such as silver halides in the photographic industry and other inorganic materials, and to organic systems such as latexes, dyes, and pigments. Other applications are for crystalline materials used as pharmaceuticals, catalysts, and imaging systems for separations and surface modifications.

Because this work is at the cutting edge of crystallization science and technology, this information has not been available from previous textbooks and academic institutions. Thus, the book provides a unique opportunity to learn up-to-date principles for precision controlled precipitations.

This practical information is aimed at chemical engineers, chemists, and other scientists who need to control precipitation processes for size control. The concepts have been used by those who are involved in the design of crystallization processes for research, quality control, product development, production processes, pilot plant operations, and manufacturing. The models have provided practical guidance for robustness in manufacturing. A deliberate effort was made to derive straightforward equations for understanding and application where basic knowledge of physical chemistry, chemical engineering, and some knowledge of calculus and of process fundamentals will be sufficient.

The benefits of this book will add to the understanding of the basic and practical principles that control crystal size in precipitations. It teaches advanced and new principles to solve precipitation problems in batch and continuous processes. It gives the principles and tools to quantitatively relate the crystal size to precipitation variables. As a special benefit, the information will teach how to minimize the number of experiments in precipitation R&D and product development. As a result of the direct correlation with manufacturing control variables, the reader will learn to predict and correct process limitations and breakdowns.

Acknowledgments

I would like to thank my wife, Helga, whose patience and tender loving care supported me through the many hours of writing this book. I am grateful for the support of my children, who patiently suffered through the times when I was excitedly recounting the progress of the manuscript when I should have paid attention to their lives. The germ of the BNG model was due to the close cooperation with my colleagues Jong Wey and Ramesh Jagannathan, who coauthored the first publication of this work. We shared the unanticipated journal award, honorary mention, for the year when this work was published. They continued with other directions in their research and left the way open for me to update and further develop the many aspects of the theory. They graciously continued to evaluate the manuscripts before submission to the publishers. I would like to thank many of my supervisors, and Eastman Kodak Company, who tolerated and sometimes supported the development and publications of the theory and experiments. I am grateful for many skillful coworkers who helped with the experiments for the theory while continuing on their assigned crystallization and other work for product development. Throughout the publications, I appreciated the input of many anonymous reviewers who gave useful and helpful advice to improve the manuscripts.

This book could not have been published without the graceful acceptance of the book proposal by the publisher, CRC Press and their Taylor and Francis Group, LLC. Here, I thank Barbara Glunn, senior editor, for her support in crossing administrative hurdles, and Patricia Roberson, production coordinator, who patiently guided me through the writing and editorial process. After so many years working on this topic, many others, who are not mentioned, have contributed to developing the work that is represented in this book. I thank them all.

Ingo H. Leubner

The Author

Ingo H. Leubner received a PhD in physical chemistry from the Technical University in Munich, Germany, where he continued his studies with a post-doctoral fellowship. At Texas Christian University, Fort Worth, he held the post-doctoral position of R. Welch Fellow, studying and teaching photochemistry. From there, he accepted the position of senior research scientist at Eastman Kodak Company. His position required working in  photographic and precipitation science and product development. His work on the use of silver halides for the development of photographic films and papers led to new insights, contributions to photographic science, and models for the control of crystal nucleation. For his published work he received the Fellowship and Service Awards, and the Lieven-Gevaert medal, the highest award in photographic science, from the Society for Imaging Science and Technology. As a team leader he guided the development of commercially successful products. As a result of his research, he became an experienced author, lecturer, scientist, and technical project manager. After separation from Eastman Kodak Company, he became founder and senior scientist for Crystallization Consulting, a company specializing in consulting, modeling, and teaching of advanced models for high-precision precipitations. Throughout his career, he has given numerous seminars and presentations on the topics of his research at national and international conferences, major universities, and industries. His publications, presentations, and seminars resulted in national and international recognition. His name is listed in *American Men and Women in Science* and in *Who's Who in Science and Engineering*. He is a fellow of Sigma Xi, a member of the American Chemical Society, the Society for Imaging Science and Technology, the American Association for the Advancement of Science, the American Geographical Union, the Rochester Academy of Science, and the Rochester Professional Consultants Network.

In his spare time, he developed a new model of the life cycle of the universe, based on the recognition of the transition of pre-universal energy singularity to (crystalline) electrons and protons, which are the building stones of the universe, fundamentally a crystallization process. This led to the unanticipated discovery of the origin of gravity and its identification as anti-energy. The reversal of the process of the formation of the universe leads to its conversion from mass to photons, and thus to a loss of cohesion. Classical modeling of this process led to the identification of the famous Hubble constant as the mass-loss rate of the universe. Application of the radiative mass-loss concept to the solar system identified the weak cohesion of the planets to the sun, their dissociation from the sun since the formation of the solar system, the when and why of a wet Mars, and many other effects that are experimentally proven or verifiable. The draft for a book on this topic is in progress. The author's broad knowledge and experience combined with his intricate involvement in crystallization processes have contributed to his success in crystallization consulting.

1 Prior Models for Nucleation and Growth

THE PRIMITIVE MODEL

A primitive model for the total number of crystals formed, Z, can be derived if one assumes that in a limited nucleation time, t_0, a number of Z crystal nuclei of the (nascent) size, l_0, are formed, and k_v is the volume constant that converts the characteristic length, l_0, to crystal volume. The mass volume of crystals formed, V_z [cm³], is given by Equation (1.1).

$$V_z = Zk_v l_0^3 \tag{1.1}$$

V_z is also equal to the mass volume added during the nucleation time, t_0 (Equation [1.2]).

$$V_z = RV_m t_0 \tag{1.2}$$

Here, V_m is the molar volume (cm³/mole of the crystals). The addition rate, R (mole/s), is constant through the nucleation period; using the mass balance, V_z, between material addition rate, R, and the mass of nascent nuclei leads to Equation (1.3):

$$Z = \frac{RV_m t_0}{k_v l_0^3} \tag{1.3}$$

Generally, arbitrary assumptions are made about either the nucleation period, t_0, or the nascent crystal size, l_0. For instance, it has been assumed that the nucleation time is the time from the start of reactant addition to the time when the turbidity of the reaction mixture reaches a given density. In reality, this density only reflects the size and number of crystals when the turbidity condition is reached. Nucleation might have ended substantially earlier, and the visual appearance of turbidity does mostly indicate the growth from a much smaller nascent to a visible crystal size.

This equation does not explicitly include important reaction variables, such as temperature and solubility, which control the nucleation process. Also, the nucleation rate depends on the supersaturation during nucleation and is not a linear process as assumed for Equation (1.3).

The origin of this model is obscured in history. Other models must be used for mechanistic modeling. Better times for the nucleation time have been determined by stop-flow procedures, and high-definition laser diffraction measurements may offer another practical technology for longer nucleation rates. However, as we will show, the nucleation rate is not constant during the nucleation process (except for continuous processes).

THE TOTAL CRYSTAL NUMBER

A practical application of the mass balance of this primitive model is the calculation of the total crystal number formed in the nucleation process.

The total crystal number, Z, is generally determined by a variation of Equation (1.3). Equation (1.4) represents the mass balance from the beginning (t_0) to the end (t_e) of the precipitation.

$$Z = \frac{F_n V_m \int_{t_0}^{t_e} R dt}{k_v l_e^3} \tag{1.4}$$

The integral is the total amount (moles) of reactants added. This approach represents the opportunity to vary the reactant addition rate, R, after the end of nucleation to control the growth rate of the crystals and to control the addition time needed to achieve a desired crystal size. The final size of the crystals, l_e, is analytically determined. F_n is the fraction of reactants converted to crystalline material and may vary from zero to one. The amount of unreacted reactants may also be analytically determined at the end of precipitation.

Equation (1.4) is the standard equation for calculating the total number of crystals formed at any time during the crystallization and will be used to calculate Z in the experimental sections.

THE CLASSICAL NUCLEATION MODEL (BECKER-DOERING)

MODEL

This model is given by Equation (1.5).

$$J = J_0 \exp \frac{-16\gamma^3 V_s^2}{3k^3 T^3 (\ln S)^2} \tag{1.5}$$

Here, J is the nucleation rate (number/cm³ sec); J_0 is the nucleation constant (number/cm³ sec); γ is the surface energy (erg/cm²); V_s is the molecular volume (molar volume/Avogadro number); k is the Boltzmann constant; T is the absolute temperature (K); S is the supersaturation, $(C - C_s)$; C is the concentration; and C_s is the equilibrium concentration.

In Equation (1.5), the usual preexponential factor A was renamed J_0, since it represents the maximum nucleation rate. The ratio of J/J_0 (Equation [1.6]), which varies between zero and one, was modeled as a function of surface energy, γ, and supersaturation, S[11] (Figure 1.1).

$$\frac{J}{J_0} = \exp \frac{-16\gamma^3 V_s^2}{3k^3 T^3 (\ln S)^2} \tag{1.6}$$

For silver halide precipitations, where S is generally greater than 10^6, continuous nucleation was predicted.[1] In contrast, it was experimentally found that a finite number of crystals are obtained, which then grow after the nucleation phase.

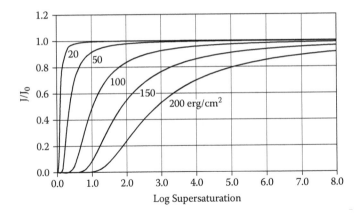

FIGURE 1.1   Relative nucleation rate, J/J_0 versus log supersaturation.

DERIVATION OF THE CLASSICAL NUCLEATION MODEL

Relatively recent derivations of this equation were given by Mutaftschiev (1993)[2] and Katz and Donohue (1979).[3] The nucleation process is modeled following the classical mechanism, which has been described in several books.[4] This model was derived for the condensation of vapor to liquid. It is generally also extended to crystallization from melts and solutions.

The model assumes that the nucleation process proceeds through various steps until the size of a stable critical cluster, $(AB)_n$, with the critical nucleus size, r_c, is reached, which has equal probability to dissolve or grow under the reaction conditions:

$$A + B \quad (AB)$$

$$AB + A + B \quad (AB)_2$$

$$(AB)_2 + A + B \quad (AB)_3 \text{ sub clusters, sub nuclei, embryos}$$

$$(AB)_{(n-1)} + A + B \rightarrow (AB)_n \text{ critical cluster/nucleus size} = r_c$$

The *critical nucleus (or cluster) size*, r_c, is the initially formed cluster that has equal probability to grow or dissolve under the given reaction conditions. The size of this nucleus is given by its thermodynamic stability (see below).

The critical nucleus size must not be confused with the *critical crystal size*, r^*, which is the crystal size where crystals have equal probability to grow or dissolve, however, *in the presence of a larger stable crystal population*. The size, r^*, is given by the reaction conditions and the size of the stable crystal population, r. The BNG model experimentally determines the ratio r^*/r. From this ratio and the knowledge of r, the supersaturation is calculated.

The size of the critical cluster is related to free energy changes associated with the process of homogeneous nucleation and may be modeled as follows: The overall excess free energy, ΔG, is equal to the sum of surface excess free energy, ΔG_s, and volume excess free energy, ΔG_v. In a supersaturated solution, ΔG_v is a negative quantity.

$$G = G_s + G_v \tag{1.7}$$

The energy terms are replaced by analytical equivalents:

$$G = 4\pi r^2 \gamma + \frac{4\pi r^3 \ G_v}{3} \tag{1.8}$$

The maximum value of ΔG, ΔG_{crit}, corresponds to the critical nucleus, r_c. For a spherical cluster, one obtains:

$$\frac{d \ G}{dr} = 8\pi r \gamma + 4\pi r^2 \ G_v = 0 \tag{1.9}$$

and

$$r_c = \frac{-2\gamma}{G_v} \tag{1.10}$$

The formation of the critical nucleus is an undisputed assumption in the classical nucleation theory. Its size is not generally experimentally or theoretically obtainable.

Solving Equation (1.10) for ΔG_v and back substituting into Equation (1.8) gives:

$$G_{crit} = \frac{4 \ \gamma r^2}{3} \tag{1.11}$$

The classical theory focuses now on the rate of nuclei formation and ignores the growth of the formed crystals. This leads to a third assumption that the nucleation rate follows the Arrhenius reaction rate equation commonly used for the rate of a thermally activated process:

$$J = J_0 \exp\left(\frac{- \ G}{kT}\right) \tag{1.12}$$

At this point, the Gibbs-Thomson relationship is introduced:

$$\ln \frac{c}{c^*} = \ln S = \frac{2\gamma V_s}{kTr} \tag{1.13}$$

which is solved to:

$$\frac{2\gamma}{r} = \frac{kT\ln S}{V_s} \tag{1.14}$$

Thus, Equations (1.15) and (1.16):

$$- G = \frac{2\gamma}{r} = \frac{kT\ln S}{V_s}$$ (1.15)

Back substitution into the Arrhenius equation (1.12) gives Equation (1.16), which is the same as Equation (1.5):

$$J = J_0 \exp \frac{-16\pi\gamma^3 V_s^2}{3k^3 T^3 (\ln S)^2}$$ (1.16)

An interesting result is obtained if J is divided by J_0:[1]

$$\frac{J}{J_0} = \exp \frac{-16\pi\gamma^3 V_s^2}{3k^3 T^3 (\ln S)^2}$$ (1.17)

We obtain for S = 1.0 (no supersaturation): $J/J_0 = 0$, and for S -> ∞ (infinite super-saturation), $J/J_0 = 1.0$.

Thus, at high supersaturation, $J = J_0$, where J_0 is the maximum nucleation rate attainable under the given reaction conditions. This condition is, for instance, present for precipitations of silver halides, where the supersaturation at the addition point of silver ions may reach orders of magnitude greater than 10^5. The critical nucleus sizes have been estimated to about four AgBr or eight AgCl clusters.[1] During the growth process, these clusters will be consumed by the growth of larger crystals.

It is important to note that growth in competition with nucleation is not included in the derivation of the classical nucleation model. The total number of crystals formed in a crystallization process is not obtainable. Further, no reference is given to experimental control parameters such as reaction addition rate, crystal solubility, and ripening and restrainer effects. The temperature effect does not relate to the total number of crystals formed. Further, the assumption that nucleation, J (No/cm³s), depends on the reaction volume has been experimentally disproved.

In industrial research, in which it is necessary to control the final crystal size and thus the nucleation process, the classical theory did not provide the needed practical guidance for precision crystallizations. This led to the development of two new approaches for modeling the crystal number as a function of reaction control variables. This will be discussed in the following sections.

THE KLEIN-MOISAR MODEL

INTRODUCTION

Klein and Moisar[5] published a nucleation model that related the total number of crystals formed, Z, to the reaction addition rate, R (mole/s), crystal solubility, C_s (mole/cm³), and reaction temperature, T (Equation [1.18]). Unfortunately, they did not provide the derivation of this equation.[5]

Subsequently, Kharitanova, Shapiro, and Bogomolov[6] and Sugimoto[7] published derivations of equations that attempted to correlate the stable number, Z, of crystals formed in controlled batch precipitations as a function of precipitation conditions. Their equations differ from the original Klein-Moisar equation by the constant, K.[6,7]

$$Z = \left[\frac{KR_g}{\gamma DV_m} \right] \frac{RT}{C_S} \tag{1.18}$$

$$K = 1.0 /8\,\pi\,(\text{Klein-Moisar}) = 3.0/8\,\pi\,(\text{Kharitanova})$$
$$= 1.0/5.9\,\pi\,(\text{Sugimoto})$$

Here, Z is the total number of stable crystals formed; R is the reactant addition rate (dmole/dt); R_g is the gas constant; D is the diffusion coefficient (cm²/s); V_m is the molar volume (cm³/mole crystal); γ is the crystal surface energy (erg/cm²); and C_s (mole/cm³) is the solution equilibrium concentration. Equation (1.18) assumes diffusion-controlled crystal growth and spherical morphology of the crystals.

This equation avoids the problems of the classical nucleation theory by not using supersaturation as a control variable.

KHARITANOVA DERIVATION

We will present the derivation by Kharitanova, Shapiro, and Bogomolov.[6,9] Their model is based on a mass balance that equates the total crystal mass and the total amount of reactants added.

Equation (1.19) gives the average crystal diameter, d, at any precipitation time. Spherical crystal morphology is assumed.

$$d = \left\{ \frac{6RV_m t}{\pi Z} \right\}^{1/3} \tag{1.19}$$

On the other hand, if Ostwald ripening of the crystals in a closed system under diffusion-controlled growth is considered, the Wagner equation (1.20) predicts an average crystal size as a function of time and other variables.[8]

$$d = \left\{ \frac{16\gamma DV_m^2 C_s t}{R_g T} \right\}^{1/3} \tag{1.20}$$

Equating Equations (1.19) and (1.20) leads to Equation (1.18) with $K = 3.0/8\pi$.

The use of the Wagner equation, which is based on a closed system, does not appear justified for crystallization systems where reactants are added throughout the nucleation and growth phase. Thus, from fundamental considerations, a more rigorous approach based on an open system is required.[5] Such an approach is given by the balanced nucleation and growth model, which will be presented in the following sections.

COMMENTS

Experimentally, we observed nonlinear correlations between the crystal number, Z, and the reaction variables R, T, and C_s. These could not be accommodated in Equation (1.18) except by the introduction of arbitrary exponents to the variables. In addition, the temperature dependence of the number of crystals is inconsistently predicted to increase with temperature or to be temperature independent, while experimentally, the crystal number generally decreases and size increases with increasing temperature during nucleation.

This problem was solved by the balanced nucleation and growth (BNG) model, which was developed by the author and coworkers.[1,9] In the BNG model, the Klein-Moisar model (Equation [1.18]) is a limiting case and thus will not be discussed further.

GROWTH RATE AND MAXIMUM GROWTH RATE

GROWTH RATE BELOW THE MAXIMUM GROWTH RATE

Before the BNG model can be derived, it is necessary to discuss the growth reaction of a crystal population during the nucleation process. For present purposes, two models are chosen, the mass-balance and the maximum growth models. For more detailed discussions of growth mechanism, other references may be used.[10] However, the current concept of maximum growth rate appears not to be presented.

The material balance of added material and consumption generally determines the growth rate. This definition is also true for the maximum growth rate (Equation [1.21]).

$$G = \frac{dr}{dt} = \left[\frac{V_m}{3.0k_v} \right] \frac{R}{r^2 Z} \tag{1.21}$$

Here, G is the growth rate, which is defined by the change of crystal size, dr/dt. R is the molar addition rate ($dmole/dt$); V_m is the molar volume ($cm^3/mole$); k_v is the crystal volume factor, which converts a characteristic crystal size into crystal volume; and Z is the number of crystals in the reactor. Since the growth rate, G, is derived from a material balance, it is independent of temperature, solubility, diffusion, convection, and other reaction conditions.

In controlled double-jet precipitations, R is given by the addition rate of the reactants. In systems where the material is formed by chemical reactions, R is given by the reaction rate. For systems where the material is provided by diffusion and convection, such as formation of droplets or in monomolecular crystallization (e.g., ice formation), R is given by the diffusion and convection rates. If temperature variation is used to control the formation of reactants, then the addition rate is given by the rate of solubility change with temperature.

MAXIMUM GROWTH RATE

Unlike the value of the growth rate, G, below maximum growth rate, the maximum growth rate, G_m, is a function of temperature, solubility, material, and the mechanism of growth, i.e., kinetically or diffusion-controlled growth. At the same time, G_m is also given by Equation (1.21).

Strong and Wey developed and experimentally supported a maximum growth model for silver halides from first principles. This model includes the effects of growth mechanism and was used to derive the general BNG model and the special cases of nucleation under diffusion and kinetically controlled growth conditions.[9–11]

Their analysis shows that the theoretical maximum growth rate, g, must be modified when both kinetically and diffusion-controlled growth are present. To obtain the modified maximum growth rate, G_m, Equation (1.22) was derived.

$$G_m = \frac{dr}{dt} = g\frac{1.0 - r^*/r}{1.0 + \varepsilon r} \tag{1.22}$$

The maximum growth rate, g, which is defined by kinetic-controlled growth, is given by Equation (1.23).

$$g = K_i(C - C_s) \tag{1.23}$$

C is the actual concentration during precipitation, and C_s is the equilibrium solubility concentration of the crystals. The difference $(C - C_s)$, the supersaturation, is obtained from the Gibbs-Thomson effect (also known as Ostwald ripening, Equation [1.24]).

$$(C - C_s) = \left[\frac{2\gamma V_m}{r^* R_g}\right]\frac{C_s}{T} \tag{1.24}$$

Stepwise back substitution of Equation (1.24) into (1.23) and (1.22) leads to the modified maximum growth rate (Equation [1.25]), which includes the ratio of kinetic- to diffusion-controlled growth ε (Equation [1.26]).

$$G_m = \left[\frac{2\gamma V_m K_i(1.0 - r^*/r)}{R_g r^*(1.0 + \varepsilon r)}\right]\frac{C_s}{T} \tag{1.25}$$

$$\varepsilon = \frac{K_i}{DV_m} \tag{1.26}$$

Here, K_i is the surface integration (reaction) constant, and ε represents the relative resistance of bulk diffusion to surface reaction. The other variables and constants were defined previously. In Equation (1.25), r is the average crystal size, and r* is the critical crystal size, which is smaller than the average crystal size, r. Crystals with the size r* have equal probability to grow or dissolve in the reaction system.

Experimentally, the maximum growth rate is determined from the transition from mass-balance growth to renucleation. An efficient way to determine G_m is described in detail in Chapter 8.

REFERENCES

1. Leubner, I. H. 1987. *J Phys Chem* 91:6069.
2. Mutaftschiev, B. 1993. *Handbook of crystal growth*. Vol. 1a. Ed. D. T. F. Hurle. Amsterdam: North-Holland.
3. Katz, J. L., and M. D. Donohue. 1979. *Adv Chem Phys* 40:137.
4. Mullin, J. W. 2000. *Crystallization*. 4th ed. Oxford: Butterworth–Heinemann.

5. Klein, E., and E. Moisar. 1963. *Ber Bunsenges Phys Chemie* 67:349.
6. Kharitanova, A. I., B. I. Shapiro, and K. S. Bogomolov. 1979. *Z Nauchn Prikl Fotogr Kinematogr* 24:34.
7. Sugimoto, T. 1990. *Proceedings of the 11th Symposium on Industrial Crystallization*, Garmisch-Partenkirchen, Germany.
8. Wagner, C. 1961. *Z. Elektrochemie* 65:581.
9. Leubner, I. H., R. Jagannathan, and J. S. Wey. 1980. *Photogr Sci Eng* 24:26.
10. Wey, J. S., and R. W. Strong. 1977. *Photogr Sci Eng* 21:248.
11. Wey, J. S., and R. W. Strong. 1979. *Photogr Sci Eng* 23:344.

2 The Balanced Nucleation and Growth Model

THE BNG MODEL AND THE NUCLEATION PHASE

The balanced nucleation and growth (BNG) model combines nucleation and growth during the nucleation phase for crystallizations. The present section is concerned with modeling of the nucleation phase, which is defined as the initial phase of the formation of stable crystals in the controlled precipitation of organic and inorganic materials.[1] After the nucleation phase, no more crystals are formed, and the formed crystals will grow. Modeled are crystal size distribution and number, supersaturation, nucleation rate, and maximum crystal population growth rate.

ASSUMPTIONS FOR MODELING

To arrive at a model that can be numerically evaluated, certain assumptions must be made. They consist of assumptions for modeling the nucleation process and for facilitating numerical evaluation.

1. During the nucleation phase, the reactant addition rate is constant.
2. The first step (time interval) consists solely of nucleation, forming a number of crystals with the critical nucleus size, r_c.
3. Until the end of nucleation, formation of new crystals and growth of existing crystals compete for the available unreacted material (supersaturation).
4. During the nucleation period, all crystals grow at maximum growth rate.
5. At the last step of nucleation, the formation of new crystals stops, and the existing crystal population is at maximum growth rate.
6. After the growth rate of the crystal population drops below the maximum growth rate, nucleation ends.

ASSUMPTIONS FOR CALCULATIONS

For convenience of calculations, the following assumptions were made. The model allows changing these assumptions.

1. The reactant addition rate is constant during the nucleation phase.
2. Chemical reactions leading to the crystallizing product follow first-order kinetics.
3. The maximum growth rate, G_m, is constant during the nucleation phase.

A similar nucleation and growth model was developed for vapor-to-liquid nucleation processes within the diffusion cloud chamber.[2,3] In that approach, both simultaneous nucleation and growth processes were considered, and the effects of both vapor depletion and latent heat were included. The classical nucleation model was used as the basis of the experiments and for the evaluation of the experimental results. The surface energy of the droplets was used as an adjustable parameter and was experimentally determined from the nucleation experiments. An important difference between the case considered here and the diffusion cloud chamber is that the latter permits a quantitative measurement of the nucleation rate.[4] This is generally not possible for liquid/solid nucleation. In the cloud chamber model, the growth process was modeled after the nucleation step, while in the present model, nucleation and growth compete simultaneously for added reactant.

Others followed our suggestion of adding a growth process.[5-8] These authors failed to include significant developments of the original model[2] in subsequent publications.[3,4] It is the author's assertion that those models are superseded by the model presented here and in previous publications.[1,12]

PRELIMINARY CONSIDERATIONS: NUCLEATION RATE

For simplicity, first-order kinetics will be assumed for the reaction rate of the rate determining reactant A, which reacts with B to the crystallizing product, AB. That is, if B is in excess, the reaction rate

$$A + B \quad AB \tag{2.1}$$

is determined by the concentration of A and the reaction conditions in the reactor. For highly insoluble and rapidly reacting materials, such as silver halides, the reaction can be controlled by the controlled addition of the reactants (silver and halide ions; controlled double-jet precipitation).

The decrease of A is given by:

$$A = A_0 \exp(-kt) \tag{2.2}$$

The formation of product AB is given by:

$$AB = A_0(1.0 - \exp(-kt)) \tag{2.3}$$

For two experiments that vary only in A_0 (A_{01} and A_{02}), the concentration of AB (AB_1 and AB_2) after a fixed time interval is given in Equation (2.4).

$$\frac{(AB)_1}{(AB)_2} = \frac{A_{01}}{A_{02}} \tag{2.4}$$

The modeling parameter, F_n, is a measure of conversion of the reaction product to crystals and is thus a measure of nucleation efficiency during a given time interval (Equation [2.5]).

$$F_n = \frac{(AB)_1}{A_{01}} = \frac{(AB)_2}{A_{02}} \tag{2.5}$$

The reaction A + B -> AB is controlled by the kinetics of materials present in the reactor.

For the nucleation reaction, the reacted material will be in the form of critical nuclei of size r_c. Since r_c is generally not known, a model parameter, L_n, was introduced to replace r_c. The number of crystals formed in a given time interval, i, is the model variable N_i.

THE BALANCED NUCLEATION AND GROWTH MODEL

MODELING VARIABLES

1. R_a, reactant addition rate, dR/dt
2. F_n, nucleation efficiency
3. L_n, r_c critical nucleus size
4. G_m, maximum growth rate
5. dt, time, nucleation interval

The balanced nucleation and growth model varies from the classical nucleation model at the point where in the classical model the Arrhenius is introduced. Instead, the BNG model is based on the assumption that the critical clusters (nuclei) will immediately after formation start growing at maximum growth rate. Due to continuous reactant addition, they will take up newly formed or newly added material. The consumption of incoming material by growth will reduce the nucleation process. In other words, nucleation will decrease as nucleation proceeds and growth increases, and eventually nucleation will be fully replaced by growth.

The overall concept for calculations is visualized in Table 2.1. The flow chart represents the procedure for the calculations and provides numerical algorithms to calculate the progress of crystal formation and growth during the nucleation period.

For the numerical evaluation of the model, the nucleation phase is broken down in time elements, dt. During the nucleation interval, dt, an amount of material, R_a, equivalent to a volume element (cm³) of crystal material is added (Equation [2.6]).

$$R_a = R \, V_m dt \tag{2.6}$$

Here, R is the molar addition rate (dmole/dt), and V_m is the molar volume (cm³/mole).

INITIAL TIME INTERVAL

During the first time interval, dt, a fraction F_n of R_a is converted into N_1 crystal nuclei of the volume, V_c (Equation [2.7]).

$$N_1 = \frac{F_n R_a dt}{V_c} \tag{2.7}$$

$$V_c = k_v r_c^3 \tag{2.8}$$

TABLE 2.1

The Balanced Nucleation and Growth (BNG) Model

Time, dt	Supersaturation, Start, SS_i	Nucleation, N_i	Growth, G_i	Supersaturation, S_i	Comments
0	0.0	0.0	0.0	0.0	0.0
1	$Ra \times dt = S1$	N1	0.0	S1	Only Nucleation
2	$Ra \times dt + S2$	N2	G(N1)	S2	Growth and Nucleation
3	$Ra \times dt + S3$	N3	G(N1) + G(N2)	S3	"
"	"	"	"	"	"
i	$Ra \times dt + Si$	Ni	Σ G(Nn) n = 1->i−1	Si	"
"	"	"	"	"	"
e	$Ra \times dt$	0.0	Σ G(Nn) n = 1-> e	0.0	End of Nucleation

Abbreviations: SS_i, starting supersaturation; S_i, ending supersaturation; N_i, number of crystals formed; G_i, crystal population maximum growth rate during a reaction interval dt = i.

The nuclei volume, V_c, is given by the volume conversion factor, k_v, and the critical nucleus size, r_c (Equation [2.8]). The magnitude of dt, the nucleation interval, affects the outcome of the calculations and should be held constant within a set of related calculations. For the present modeling, a value of 0.1 was chosen for dt.

A fraction of the added material, S_1, remains unreacted and is present at the beginning of the following time interval (Equation [2.9]).

$$S_1 = (1.0 - F_n)R_a dt \tag{2.9}$$

S_1 is referred to as *supersaturation* and generally denoted as S_i (cm^3 reaction mass), since it has the function of the classical term *supersaturation* (=mole/l). In the BNG model, the mass addition and mass reaction are modeled.

SECOND AND FOLLOWING TIME INTERVALS

At the beginning of the second time interval, S_1 (the supersaturation) is the unreacted mass left over from the first time interval. To this, another amount of material is added through the material addition, R_a. Thus, in the second interval, the amount available for growth and nucleation is given by the supersaturation (as defined above), SS_2 (Equation [2.10]).

$$SS_2 = S_1 + R_a dt \tag{2.10}$$

If one assumes first-order kinetics (see above), the nucleation efficiency F_n for this and the following intervals is the same as for the first time interval, and the number of crystals formed in this interval is (Equation [2.11]):

$$N_2 = \frac{F_n SS_2}{V_c} \tag{2.11}$$

At the same time, the crystals formed during the first time interval grow with maximum growth rate, G_m (Equation [2.12]). Here, it is assumed that the crystals have cubic shape with edge length a ($k_v= 1.0$). G_m is used as an input variable in the modeling.

$$G_m = \frac{da}{dt} \tag{2.12}$$

The volume, V_c, of the stable nuclei is calculated from the mass balance of the first nucleation interval (Equation [2.13]):

$$V_c = \frac{F_n R_a}{N_1} \tag{2.13}$$

The size of the nuclei is then given by Equation (2.14):

$$L_n = \left(\frac{V_c}{k_v}\right)^{1/3} \tag{2.14}$$

The amount of SS_2 taken up in interval $i = 2$, G_2, by the crystals, N_1, which formed in the first interval, is given by Equation (2.15). In our calculations, we assume cubic morphology of the crystals, and G_m is the growth rate of each surface. Thus, the edge length increases by $2G_m dt$ at each time interval. For cubic morphology, k_v is equal to one.

Equation (2.15) can be reduced to Equation (2.16) by mathematical solution of the content in the outer brackets.

$$G_2 = N_1 k_v \left\{ (L_n + 2G_m dt)^3 - L_n^3 \right\} \tag{2.15}$$

$$G_2 = N_1 k_v \left\{ 3(2G_m)dt L_n^2 \right\} \tag{2.16}$$

For other than cubic morphology, the maximum growth rate must be adjusted for each surface area of the crystals and summarized at each interval to obtain the overall growth rate. Similarly, the volume constant, k_v, must be adjusted according to the crystal morphology. For spheres, k_v equals $4\pi/3$, and the crystal growth per time interval is $(r + G_m dt)$.

The growth rate shown in Equation (2.16) was rewritten to calculate the amount of molar addition rate, R_c, taken up by a crystal population, N_1, of the size, r_c, under conditions of maximum growth rate, G_m (Equation [2.17]):

$$R_c V_m dt = 3.0 N_1 k_v G_m dt\, r_c^2 \tag{2.17}$$

$(R_a - R_c)$ is the amount of input material available for nucleation.

To simplify the numerical treatment, G_m, R_a, and F_n were held constant through the nucleation/growth period. However, variations of these factors during the nucleation phase can easily be varied in the calculations if this is desired.

SUBSEQUENT TIME INTERVALS

In successive time intervals, the procedure used in the second time interval is extended. Maximum growth of the crystals formed in the previous time intervals continues. Nucleation is determined by the supersaturation at the beginning of the following time interval.

$$(R_a + SS_{i-1}) dt = \sum_{1}^{i-1} 3.0\, k_v G_m N_{(i-1)} r_{(i-1)}^2 dt + N_i V_c + S_i \tag{2.18}$$

Here, the sum term is the growth material uptake of the crystals formed in the intervals 1 -> i−1.

END OF NUCLEATION, FINAL TIME INTERVAL FOR NUCLEATION, T_E

During the last interval of nucleation, the formation of crystals ends, and the existing crystals grow at maximum growth rate, consuming all incoming material for growth. At this point, the supersaturation, S_i, and nucleation number, N_i, become zero (Equation [2.19]):

$$R_a = \sum_{1}^{e-1} 3.0\, k_v G_m N_{(e-1)} r_{(e-1)}^2 dt \tag{2.19}$$

To facilitate further modeling of the BNG model, the sum is replaced by the integral, and (N_{e-1}) is replaced by crystal population, n, with the size of r and the size range, dr. Further, R_a is replaced using Equation (2.6), and the equation is divided by dt on both sides (Equation [2.20]).

$$R\, V_m = k_v G_m \int_{r=0}^{\yen} r^2 n\, dr \tag{2.20}$$

This equation will be used as a starting point in Chapter 3 to derive the number of stable crystals formed under diffusion- and kinetically controlled growth conditions.

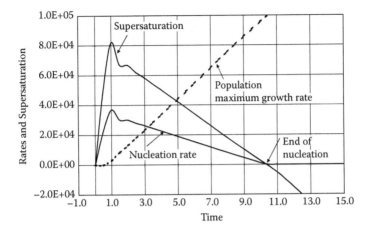

FIGURE 2.1 Nucleation rate, supersaturation, and population growth rate vs. time.

An example for the calculation of the nucleation rate, supersaturation, and population maximum growth rate is shown in Figure 2.1.

CRYSTAL SIZE AND SIZE DISTRIBUTION

With the knowledge of the number of crystals formed at a given time during nucleation, the assumption of maximum growth rate, and the calculated time of nucleation, it is straightforward to calculate the crystal size of the crystals formed at the end of nucleation.[5] This gives the crystal size distribution, L_t, and the maximum crystal size, L_m. The calculation of crystal sizes and size distribution is given by Equation (2.21):

$$L_t = L_n + (t_e - t_n)*G_m \qquad (2.21)$$

Here, t_n is the time of crystal formation, L_n is the size of the crystal at formation (critical nucleus size), t_e is the nucleation time, and G_m is the maximum crystal growth rate. For the present calculations, it is assumed that G_m is size independent.

The frame has now been put together to explore the effect of varying R_a, L_n, F_n, and G_m on the final size distribution and maximum size, nucleation rate, supersaturation, and crystal population maximum growth rate, which will be shown in the following section.

BNG MODELING OF THE NUCLEATION PHASE AND SIZE DISTRIBUTION

In the previous section, the BNG model for the nucleation phase was developed. In this section, the parameters of the BNG model of the nucleation phase are evaluated for their effect on the nucleation phase and the resulting crystal population.

In particular, it is the aim to model the nucleation rate and supersaturation as a function of reaction time. Further, as shown in the previous section, the modeling of the nucleation phase leads to the properties of the crystal population, such as size distribution, size at the maximum of crystal distribution ("maximum crystal size"), and total crystal number and the population ("maximum crystal number"). Let us review the parameters that were chosen for the calculations.

REACTION PARAMETERS FOR THE CALCULATIONS

Any of the parameters may be varied during the nucleation phase, either by calculation or, for long nucleation phases, during the nucleation phase. For the present modeling, however, they were held constant for showing their main effects on nucleation and resultant crystal population.

1. G_m, the maximum growth rate, was held constant throughout the nucleation time. The maximum growth rate may be a function of crystal size. If known, this dependence may be added to the calculations. G_m is also dependent on reaction conditions.
2. R_a, the molar addition rate, is held constant throughout nucleation.
3. F_n, the nucleation efficiency, is the fraction of material converted to nuclei.
4. L_n (r_c) is the variable used to represent the critical nucleus size.

The nucleation interval, dt, was held constant throughout the calculations. The magnitude of the nucleation interval, dt, as expected, affects the calculation of the nucleation phase and crystal population. Initial calculations show that smaller values of dt shorten the nucleation phase but provide more data points than larger values of dt. The trends shown in the following are not changed.

N_i is the number of crystals formed in a time interval. L_n (r_c) and N_i are correlated by Equation (2.8) and can thus be used interchangeably for modeling.

$$N_i = \frac{F_n(SS_i + R_a * dt)}{k_v r_c^3} \qquad (2.22)$$

For the present modeling, L_n was used instead of the critical nucleus size, r_c, to indicate that L_n is a modeling parameter.

RESULTS AND DISCUSSION

TIME DEPENDENCE OF NUCLEATION RATE, GROWTH RATE, AND SUPERSATURATION

An example of a model calculation for the nucleation phase is shown in Figure 2.2. The standard modeling parameters were used (Table 2.2).

Figure 2.2 shows the modeling results of the key processes during the nucleation phase, supersaturation, nucleation rate, and population maximum growth rate. By definition, changes of nucleation rate and supersaturation are related. Both nucleation rate and the supersaturation initially increase and then decrease. At the end of

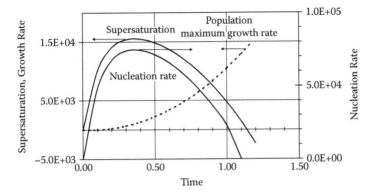

FIGURE 2.2 The nucleation phase.

the nucleation time, t_e, both become zero. The supersaturation continues to decrease below the critical supersaturation.

Negative values of the supersaturation (relative to critical supersaturation) represent the resistance of the system to renucleation, that is, the formation of new crystals. The maximum growth rate of the crystal population continues to increase steadily beyond the end of nucleation.

Size Distribution

Figure 2.3 shows an example for the calculation of the size distribution. The same model parameters were used as for the calculation of the nucleation phase (Table 2.2). For ease of comparison, the crystal numbers for the different crystal sizes were normalized relative to the maximum crystal number (100%). The plot shows that the modeling nucleation phase predicts the size distribution, the range of crystal sizes, and the size with the maximum crystal number ("maximum crystal size"). This prediction is in agreement with experimental observations.

In the following it will be explored how the various reaction parameters affect the nucleation rate and the crystal size parameters.

TABLE 2.2

Standard Values and Ranges of Modeling Parameters

Variable	Standard Value	Value Range	
R_a	10^5	$0.5–2.0 \times 10^5$	Addition rate
G_m	0.06	0.0–0.1	Maximum growth rate
L_n	0.50	0.25–1.0	Model nucleus size
F_n	0.60	0.10–1.00	Nucleation efficiency
dt	0.10	0.10	Nucleation interval

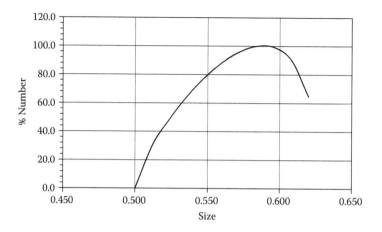

FIGURE 2.3   Crystal size distribution.

MODELING THE REACTION PARAMETERS

The variables and their modeling ranges for the calculations are listed in Table 2.2. For each model parameter, the variation of nucleation rate with time, the size distribution, and the correlation of crystal size and maximum nucleation number will be shown separately in Figures 2.4–2.6. Key results are nucleation time, t_e, maximum crystal size, and the maximum nucleation number, N_m. The maximum crystal size, L_m, is the size where the highest population density is obtained. The maximum nucleation number, N_m, is the calculated nuclei number at this size.

Addition Rate, R_a

The addition rate, R_a, was varied from 0.5 to 2.0×10^5 while the other parameters, critical nucleation size, L_n, the nucleation efficiency, F_n, and the maximum growth rate, G_m, were at the standard condition listed in Table 2.2.

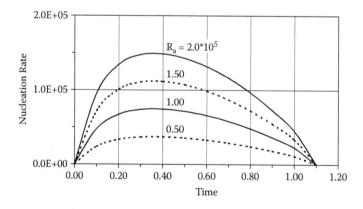

FIGURE 2.4   Nucleation rate as a function of addition rate, R_a.

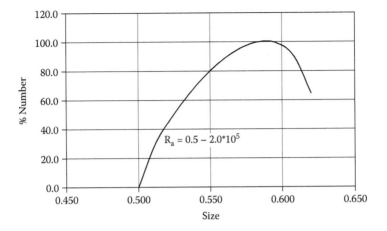

FIGURE 2.5   Size distribution as a function of addition rate, R_a.

Figure 2.4 shows that the nucleation rate increases with increasing addition rate, R_a, and that the same nucleation time, t_e, is obtained for all conditions. Figure 2.5 shows that the size and size distribution are independent of addition rate. Figure 2.6 shows that the crystal number increases linearly with addition rate, while the crystal size stays constant. This pattern is characteristic for the effect of reactant addition rate on experimental results of crystallizations.

Maximum Growth Rate, G_m

The model parameter for maximum growth rate, G_m, was varied from zero to 0.10. The other parameters, addition rate, R_a, critical nucleation size, L_n, and the nucleation efficiency, F_n, were at the standard conditions listed in Table 2.2.

Figure 2.7 shows that for zero growth rates, the nucleation rate becomes constant after a short induction period. Growth rates above zero lead to limited nucleation

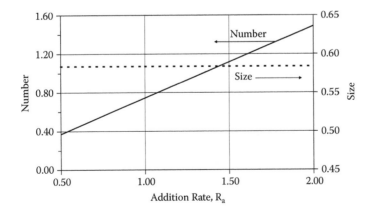

FIGURE 2.6   Crystal size and number as a function of addition rate, R_a.

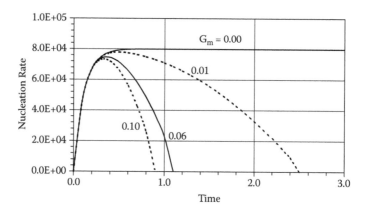

FIGURE 2.7   Nucleation rate as a function of growth rate, G_m.

periods, which shorten with increasing growth rates. This leads to narrowing of the size distribution and smaller crystal sizes. For zero growth rate, the size distribution is zero and the size is equal to the critical nucleus size (Figure 2.8). Figure 2.9 shows that the crystal number decreases, and the average crystal size increases nonlinearly with growth rate.

Critical Nucleus Size, L_n

The critical nucleus size, L_n, molar addition rate, and nucleation efficiency determine the number of critical nuclei formed. Thus, the effect of L_n on the nucleation phase was modeled in Figures 2.10 and 2.11. The molar addition rate was held constant at 1×10^5, the nucleation efficiency, F_n, at 0.6, and the growth rate, G_m, at 0.06.

With increasing nucleus size, L_n, the nucleation rate decreases and the nucleation time increases (Figure 2.10). The crystal size increases with increasing critical

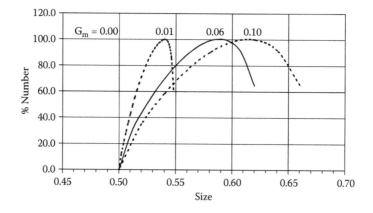

FIGURE 2.8   Size distribution as a function of growth rate, G_m.

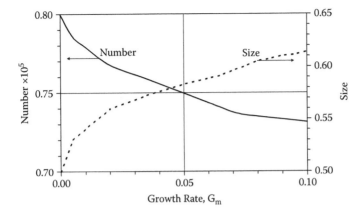

FIGURE 2.9 Crystal size and number as a function of growth rate, G_m.

nucleus size (Figure 2.11). Figure 2.12 shows that the crystal size increases linearly with critical nucleus size, while the crystal number decreases nonlinearly.

Nucleation Efficiency, F_n

The nucleation efficiency, F_n, was chosen to 0.1, 0.6, and 1.0. The addition rate, R_a, maximum growth rate, G_m, and critical nucleus size, L_n, were held at the standard conditions.

Figures 2.13 to 2.15 reveal interesting patterns of nucleation and supersaturation as a function of nucleation efficiency. Figure 2.13 shows that the nucleation rate decreases with decreasing nucleation efficiency and that the nucleation period increases. This leads to increasing crystal size and size distribution as the nucleation efficiency decreases (Figure 2.14). Figure 2.15 shows that the crystal number increases linearly with nucleation efficiency. At the same time, the crystal number decreases nonlinearly.

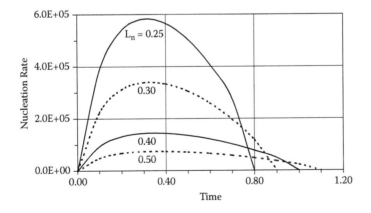

FIGURE 2.10 Nucleation rate as a function of nucleus size, L_n.

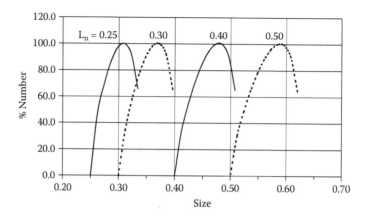

FIGURE 2.11   Size distribution as a function of nucleus size, L_n.

Practical Considerations

What does the model suggest for experimental modifications for the outcome of the nucleation process?

Nucleation (L_n), growth processes (G_m), and nucleation efficiency (F_n) control the crystal number and size distribution. These control factors are determined by the molecular and crystal properties of the crystals formed, reaction environment, and reaction conditions. They also control the growth mechanism and thus morphology of the crystals.

The reactant addition rate, R_a, is a process variable that is independent of the others. For practical considerations, it is important that the model predicts that the average crystal size and the size distribution are not affected by changing addition rate, R_a. However, the crystal number is predicted to increase linearly with R_a. Since R_a is not expected to affect the reaction processes, its variation is anticipated to allow

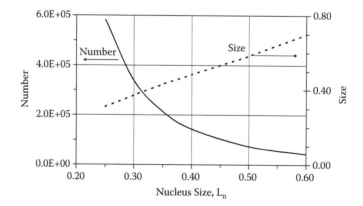

FIGURE 2.12   Crystal size and number as a function of nucleus size, L_n.

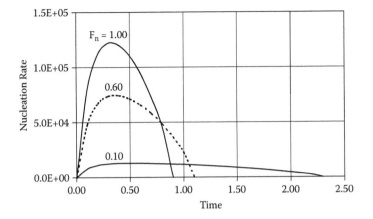

FIGURE 2.13 Nucleation rate as a function of nucleation efficiency, F_n.

practical scaling of reactions without affecting the population size distribution and crystal morphology.

These modeling results, as well as invariant experimental results, are based on the assumption that mixing of materials and uniform stirring of the reaction mixture are maintained during the nucleation process. Nucleation processes proceed during reaction conditions that sustain critical supersaturation, and thus they are anticipated to be independent of reaction volume. Reaction volume is, however, important before critical supersaturation is obtained, since it requires material that will not be available for nucleation and growth. This will be experimentally supported in the following chapters.

An important result of this model is that the growth process will codetermine the outcome of the nucleation process. Thus, different nucleation mechanisms and different

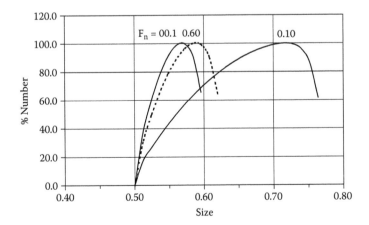

FIGURE 2.14 Size distribution as a function of nucleation efficiency, F_n.

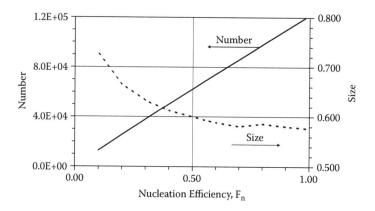

FIGURE 2.15   Crystal size and number as a function of nucleation efficiency, F_n.

experimental results are expected if the growth is either diffusion or kinetically controlled. Similarly, materials that adsorb on the crystal surface reduce the crystal growth rate. According to the present model, this enhances the nucleation process and is anticipated to result in an increase in the number of stable crystals at the end of the nucleation process. This will be modeled and experimentally supported in Chapter 7.

The BNG model predicts that if the growth process is above the maximum growth rate of the system, new crystals are formed to convert excess supersaturation into solid material. This link between growth rate and nucleation was also modeled for seeded systems, which represent a case of heterogeneous nucleation. This application of the BNG model results in an improved experimental process for determining the maximum growth rate of crystals with improved accuracy.

The BNG process applies to batch processes, as discussed here, but was anticipated to also apply for continuous precipitation processes. This was confirmed for precipitations in the controlled continuous stirred tank reactor. New equations to describe the continuous crystallization process were obtained and experimentally confirmed.

CONCLUSION

The nucleation/growth phase of the balanced nucleation and growth process was modeled to predict crystal size and size distribution as a function of four model parameters. Nucleation rate, supersaturation, and population maximum growth rate were modeled as a function of time. The adjustable model parameters were molar addition rate, R_a, critical crystal size, L_n (r_c), nucleation efficiency, F_n, and growth rate, G_m.

The model correctly describes experimental results showing that many crystallization processes lead to a limited number of crystals during a nucleation period followed by growth. The present model adds new insights to the processes that affect the transient phase of the nucleation process and the properties of the resulting crystal population.

REFERENCES

1. Leubner, I. H. 2001. A balanced nucleation and growth model for controlled precipitations. *J Disp Sci Technol* 22:125–38.
2. Brito, J., and R. H. Heist. 1982. *Chem Eng Commun* 15:133.
3. Heist, R. H., A. Kacker, and J. Brito. 1984. *Chem Eng Commun* 28:117.
4. Heist, R. H., and A. Kacker. 1985. *J Chem Phys* 82:2734.
5. Muhr, H., R. David, J. Villermaux, and P. H. Jezequel. 1995. Crystallization and precipitation engineering. Pt. 5. Simulation of the precipitation of silver bromide octahedral crystals in a double-jet semi-batch reactor. *Chem Eng Sci* 50:345–55.
6. Ludwig, F. P., and J. Schmelzer. 1995. Cluster formation and growth in segregation processes with constant rates of supply of monomers. *Zeitschrift für Physikalische Chemie* 192:155–67.
7. Muhr, H., R. David, J. Villermaux, and P. H. Jezequel. 1996. Crystallization and precipitation engineering. Pt. 6. Solving population balance in the case of the precipitation of silver bromide crystals with high primary nucleation rates by using the first order upwind differentiation. *Chem Eng Sci* 51:309–19.
8. Kresta, S. M., G. L. Anthieren, and K. Parsiegla. 2004. Mixing effects in silver halide precipitations: Linking theory with practice using a multi-mechanism model. *Chem Eng Res Des* 82 (9): 1117–36.
9. Leubner, I. H., R. Jagannathan, and J. S. Wey. 1980. Formation of silver bromide crystals in double-jet precipitation. *Photogr Sci Eng* 24:268–72.
10. Leubner, I. H. 1985. Formation of silver halide crystals in double-jet precipitations: AgCl. *J Imaging Sci* 29:219–25.
11. Leubner, I. H. 1987. Crystal formation (nucleation) under kinetically controlled and diffusion-controlled growth conditions. *J Phys Chem* 91:6069–73.
12. Leubner, I. H. 2002. Balanced nucleation and growth model for controlled crystal size distribution. *J Disp Sci Tech* 23:577.

3 The Analytical BNG Model

CRYSTAL FORMATION (NUCLEATION) UNDER KINETICALLY AND DIFFUSION-CONTROLLED GROWTH CONDITIONS

DERIVATION OF THE GENERAL BNG NUCLEATION EQUATION

In Chapter 2, the nucleation phase was modeled as a function of addition rate, R_a, maximum growth rate, G_m, critical crystal size, L_n, and nucleation efficiency, F_n. The results provided information for these factors on the course of the nucleation rate and supersaturation, and on the size distribution, maximum crystal size, and number of crystals.

$$G_m = \left[\frac{2\gamma V_m K_i (1.0 - r^*/r)}{R_g r^* (1.0 + \varepsilon r)} \right] \frac{C_s}{T} \tag{3.1}$$

$$\varepsilon = \frac{K_i}{DV_m}, \text{or:} K_i = \varepsilon DV_m \tag{3.2}$$

This evaluation, however, does not provide information on the effect of practical control parameters of crystallizations on the crystal size and number after nucleation.

The crystal size, however, is a composite of the number of crystals formed, i.e., nucleation, and growth due to the total amount of added material converted to crystals. Of these, nucleation is the critical first step of crystallization. Thus, it is desirable to relate the total number of crystals formed to actual precipitation control factors.

In this section, we will derive equations that correlate the final crystal number with the experimental addition rate, crystal solubility, and reaction temperature. This will be achieved by the introduction of the equation of the maximum growth rate, Equation (3.1), into the nucleation model (Equation [3.3]).

$$RV_m = \int_0^\infty k_v G_m r^2 n \, dr \tag{3.3}$$

For the derivation of the general nucleation model, K_i from Equation (3.2) is inserted into Equation (3.1), and the modified growth equation is inserted in Equation (3.3). After further standard manipulations, Equation (3.4) is obtained.[1]

$$R = \left[\frac{(2k_v \gamma DV_m)}{R_g} \right] \frac{C_s}{T} \left[\int_0^\infty \frac{(r^2/r^*)n \, dr}{(r + 1/\varepsilon)} - \frac{rn \, dr}{(r + 1/\varepsilon)} \right] \tag{3.4}$$

The crystal number is implicit in the integrals. Equation (3.4) describes the nucleation for the diffusion- and kinetically controlled crystal growth conditions. Since Equation (3.4) is difficult to evaluate, we will concentrate on the two extreme cases, first diffusion-controlled and then kinetically controlled growth conditions during the nucleation phase.

NUCLEATION UNDER DIFFUSION-CONTROLLED GROWTH CONDITIONS

For diffusion-controlled growth, e.g., AgCl and AgBr, the diffusion coefficient is significantly smaller than the kinetic integration constant, and thus ε is relatively large. If we introduce the approximation of Equation (3.5), Equation (3.4) reduces to Equation (3.6).[2]

$$(r + 1/\varepsilon) \gg r \tag{3.5}$$

$$R = \left[\frac{(2k_v \gamma D V_m)}{R_g} \right] \frac{C_s}{T} \int_0^\infty (rn \, dr/r^* - n \, dr) \tag{3.6}$$

The values for the integrals in Equation (3.6) are replaced as indicated in Equations (3.7) and (3.8). The critical crystal size, r*, is a constant under defined crystallization conditions.

$$Z = \int_0^\infty n \, dr \tag{3.7}$$

$$Zr = \int_0^\infty nr \, dr \tag{3.8}$$

Z is the total crystal number, and in Equation (3.8), the crystal size, r, on the left side represents the average number-weighted crystal size.

This leads to Equation (3.9), which relates the number of crystals formed under diffusion-controlled growth conditions to the fundamental experimental control parameters, reactant addition rate, R, crystal solubility, C_s, and temperature, T. The values in the square brackets are constants that are determined by independent physicochemical measurements. The value of r/r* is determined from the experiment, where r* is used to calculate to the supersaturation. Since all values in the equation are defined unambiguously, no arbitrary adjustable parameters, also known as fudge factors, are present.

$$Z = \left[\frac{R_g}{2k_v \gamma D V_m} \right] \frac{RT}{C_s (r/r^* - 1.0)} \tag{3.9}$$

Equation (3.9) describes well the nucleation of AgBr and AgCl in controlled double-jet precipitations.[2,3]

NUCLEATION UNDER KINETICALLY CONTROLLED GROWTH CONDITIONS

Under kinetically controlled growth conditions, the diffusion of material to the crystal surface is significantly faster than its integration into the surface, i.e., $D \gg K$. In this case, ε becomes smaller than 1.0, and we can introduce Equation (3.10) as an approximation.[1]

$$r + 1/\varepsilon \gg 1/\varepsilon \qquad (3.10)$$

This leads to Equation (3.11):

$$R = \left[\frac{(2k_v \gamma D V_m)}{R_g} \right] \frac{C_s}{T} \frac{\varepsilon}{r*} \left[\int_0^\infty r^2 n \, dr - rn \, dr \right] \qquad (3.11)$$

The second part of the integral was defined in Equation (3.8), which leaves to define the first part. This part of the integral presents the total surface area of the system and can be expressed by the total crystal number, Z, and the area-weighted crystal size, r_a (Equation [3.12]).

$$Z r_a^2 = \int_0^\infty r^2 n \, dr \qquad (3.12)$$

After inserting Equations (3.8) and (3.12) into (3.11), we obtain the dependence of the crystal number on the experimental parameters for kinetically controlled growth conditions during nucleation (Equation [3.13]).

$$Z = \left[\frac{R_g}{2k_v \gamma K_i} \right] \frac{RT}{C_s (r_a^2/r* - r)} \qquad (3.13)$$

To evaluate this equation, the number- and area-weighted crystal sizes, r and r_a, must be determined. For relative mono-disperse emulsions, which are generally obtained in controlled double-jet precipitations, we can use the approximation $r_a \sim r$. This leads to Equation (3.14).

$$Z r = \left[\frac{R_g}{2k_v \gamma K_i} \right] \frac{RT}{C_s (r/r* - 1.0)} \qquad (3.14)$$

This equation leads to surprising predictions:

1. Nucleation under kinetically and diffusion-controlled conditions is controlled by the same control parameters, R, T, and C_s.
2. These control parameters provide the same fundamental correlation and thus cannot be used to discern between kinetically and diffusion-controlled nucleation.
3. Intrinsic differences between diffusion- and kinetically controlled conditions are the following:
   a. For diffusion-controlled nucleation, the crystal number is constant after nucleation (Equation [3.9]).
   b. For kinetically controlled nucleation, the number of crystals decreases with size increase, so that $Z \times r$ is constant (Equation [3.14]).

The last prediction is fully unexpected and has not been reported. Its proof will be an important step to support the BNG model. An example for this prediction will be shown in Chapter 4.

REFERENCES

1. Leubner, I. H. 1987. Crystal formation (nucleation) under kinetically controlled and diffusion controlled growth conditions. *J Phys Chem* 91:6069–73.
2. Leubner, I. H., R. Jagannathan, and J. S. Wey. 1980. Formation of silver bromide crystals in double-jet precipitations. *Photogr Sci Eng* 24:268–72.
3. Leubner, I. H. 1985. Formation of silver halide crystals in double-jet precipitations: AgCl. *J Imaging Sci* 26:219–25.

4 Kinetically Controlled Nucleation

EXPERIMENT: KINETICALLY CONTROLLED CRYSTAL NUCLEATION

THEORY

In Chapter 3, Equation (4.1) was derived to correlate the number of crystals formed under kinetically controlled experimental conditions with the experimental control parameters.

$$Zr = \left[\frac{R_g}{2k_v \gamma K_i} \right] \frac{RT}{C_s (r/r^* - 1.0)} \tag{4.1}$$

This equation predicts that the number of crystals should decrease as the crystal size increases, as long as kinetically controlled growth conditions dominate.

A decrease in crystal number over an extended time of precipitation was reported by Daubendiek for AgI under certain precipitation conditions.[1] The precipitation conditions were given as pAg 8.0 and 35°C. The pAg is a measurement of the concentration of free silver ions in the reaction mixture and is defined analog to pH as $-\log[Ag^+]$, where $[Ag^+]$ is the silver ion activity. The progress of the crystallization was given by the moles of AgI precipitated, which is equivalent to time. The crystal size (r) was given as crystal linear dimension (CLD); a relative crystal number, Z, was calculated by dividing the precipitate AgI volume (moles AgI × molar volume) by CLD.[3] The experimental conditions are given in Appendix I.[1] The reported results are listed in Table 4.1.

The data show that the crystal size, r, steadily increases throughout the experiment. However, it also shows that the crystal number decreases from Experiments 1 to 4. For Experiments 4 through 6, the crystal number is constant.

As suggested by the theory, the crystal size, r, and number, Z, were multiplied, and Z × r is listed in Table 4.1. As predicted by the theory, Z × r is constant from Experiments 1 through 4, although the crystal number decreases by a factor of three. For Experiments 5 and 6, Z × r is significantly higher. This is in agreement with a transition from kinetically to diffusion-controlled growth processes between Experiments 4 and 5. Crystal number and Z × r are also plotted in Figure 4.1.

The figure clearly shows the decline in crystal number and the constant value or Z × r during the initial phase of the precipitation.

TABLE 4.1

Effect of Precipitation Time on the Crystal Size and Number of an AgI Precipitation

Exp.#	Moles of AgI Formed	r, nm	$Z10^{-7}$	$Z \times r\ 10^{-10}$
1	0.255	45	26.8	1.2
2	0.860	81	12.6	1.1
3	1.37	107	11.0	1.2
4	2.04	129	9.4	1.2
5	2.91	147	9.5	1.4
6	4.00	166	9.3	1.5

Size (r) is a crystal linear dimension (nm, CLD); Z is a relative crystal number; precipitation conditions: pAg 8.0, 35°C (Appendix I)

DETERMINATION OF THE KINETIC INTEGRATION CONSTANT, K_i

The data show that the crystal number initially decreases, indicating kinetically controlled nucleation, and that around 2.0 moles AgI the crystal number becomes constant, indicating diffusion-controlled growth. This transition is due to the increase of the surface area of the crystal population beyond the critical area where the rate reactant addition decreases below the rate necessary to control crystal growth. Instead, the addition rate is low enough that material diffusion controls the growth process. The transition from kinetically to diffusion-controlled growth conditions allows determining the kinetic integration constant, K_i.

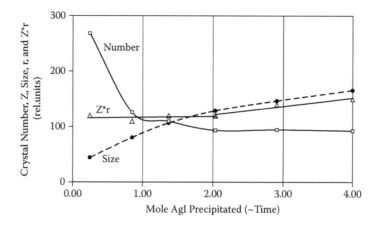

FIGURE 4.1   Kinetically controlled nucleation in a silver iodide precipitation.

At the critical condition where this transition occurs, both nucleation equations apply, so that the equation for diffusion-controlled conditions (Equation [4.2]) can be set equal to the equation for kinetically controlled conditions (Equation [4.1]).

$$Z = \left[\frac{R_g}{2k_v \gamma DV_m} \right] \frac{RT}{C_s(r/r^* - 1.0)} \tag{4.2}$$

This leads to Equation (4.3), where r_t is the crystal size at the transition condition.

$$K_i = \frac{DV_m}{r_t} \tag{4.3}$$

We can now estimate the magnitude of K_i. We estimate from Table 4.1 that r_t is about 65 nm (CLD/2). For 35°C, the diffusion constant, D, was estimated[2] to 0.88E-05, and the molar volume of the hexagonal structure of AgI is 41.4 cm^3/mole. These data give an estimate of K_i of about 56 cm^4/(s mole). It is believed that this is the first time that the kinetic integration constant could be determined from precipitation experiments.

MECHANISM

The mechanism of the decrease of crystal number under kinetically controlled conditions may be explained by the physical processes occurring during growth.

For diffusion-controlled conditions, the kinetic integration rate is greater than the diffusion process, and all incoming material is integrated into the crystal surface.

Under kinetically controlled growth conditions, the diffusion rate to the crystal is in excess of the rate of material incorporation. Thus, growth material accumulates at the crystal surface. In the case of iodide, this leads to the formation of AgI_n^{1-n} species. These species are highly soluble and diffuse material away from the surface. This material can be deposited on larger crystals by Ostwald ripening. At the same time, small crystals will shrink below the critical crystal size and dissolve. It is an unanticipated prediction and experimental fact that growth and loss of crystals are related.

CONCLUSION

In conclusion, the surprising prediction of the BNG theory (Equation [4.1]) was confirmed: Under kinetically controlled growth condition, the number of crystals should decrease inversely proportional to crystal size. The product Z x r was found to be constant over an extended time for an AgI precipitation under specific conditions. In addition, it was possible to calculate the kinetic integration constant from the transition conditions from kinetically to diffusion-controlled growth. The experimental support of the theoretical prediction provides significant support to the BNG model.

APPENDIX I: EXPERIMENTAL CONDITIONS
FOR AgI PRECIPITATION

All precipitations were done at 35°C, using a double-jet-controlled pAg method.[1,3,4] The pAg is a measurement of the concentration of free silver ions in the reaction mixture and is defined analog to pH as $-\log[Ag^+]$, where $[Ag^+]$ is the silver ion activity. Apparatus and mode of reactant introduction and pAg control were similar to those previously described.[5] Reactants were $AgNO_3$ and NaI, which were added at controlled rates to an agitated solution of lime-processed gelatin in water. The silver ion potential was monitored with an AgI-plated silver electrode relative to an Ag/AgCl/4.0 N KCl reference electrode. The electrode system was calibrated prior to each experiment with dilute NaI or $AgNO_3$. Temperature and pAg were held constant throughout the experiments. Size increases during preparation and storage due to Ostwald ripening were minimized by adjusting to the minimum of solubility at pI5 $(-\log[I-])$.

The crystal size (mean CLD, crystal linear dimension) and size distribution were determined from direct transmission electron micrographs. Generally, 300 particles were sized for each mean CLD determination and size distribution measurement. For triangular crystals, the length of the side was taken for CLD; for hexagonal bipyramids, the distance between opposite basal corners was chosen; and for irregular shapes, equivalent circular diameters were recorded. The size distribution is given as one standard deviation.

REFERENCES

1. Daubendiek, R. L. 1978. *Proc Int Congr Photogr Sci* 141.
2. Leubner, I. H. 1985. *J Imaging Sci* 29:219.
3. Klein, E., and E. Moisar. 1963. *Ber Bunsenges Phys Chem* 67:349.
4. Berry, C. R., and D. C. Skillman. *J Phys Chem* 68:1138; Berriman, R. W. 1964. *J Photogr Sci* 12:121.
5. Wey, J. S., and R. W. Strong. 1977. *Photogr Sci Eng* 21:14.

5 Diffusion-Controlled Nucleation

REACTION VARIABLES, CONSTANTS, PREDICTIONS

THEORY

In Chapter 4, Equation (5.1) was derived to correlate the number of crystals formed under diffusion-controlled experimental conditions with experimental control parameters.

$$Z = \left[\frac{R_g}{2k_v \gamma D V_m} \right] \frac{RT}{C_s (r/r^* - 1.0)} \tag{5.1}$$

Here, Z is the total number of crystals formed. The reaction control parameters are R, addition rate (mole/min); T, temperature (K); and C_s (mole/l), the solubility of the reaction-controlling reactant.

CRITICAL CRYSTAL SIZE, r*, AND SUPERSATURATION

The following constants are independently determined. R_g is the general gas constant, γ is the surface energy (erg/cm^2), D is the diffusion constant (cm^2/s), V_m is the crystal molar volume (cm^3/mole), and k_v is the crystal volume constant that converts the size measurement into volume.

In the equation, r is the average, usually number-weighted, crystal size. The critical crystal size, r*, is the crystal size that under the given reaction conditions and in the presence of the average crystal size, r, has equal probability to dissolve or grow.

Under diffusion-controlled nucleation conditions, Z is a constant, and thus r/r* is also a constant, which can be determined if all other factors in Equation (5.1) are known. With the knowledge of the average crystal size and r/r*, r* can be determined. Using the Kelvin/Thomson/Ostwald equation, (5.2), the supersaturation, S, of the reaction mixture can be calculated for any time during the precipitation as long as the crystal number is constant.

$$S = \frac{C^*}{C_s} = 1.0 + \frac{2.0 \gamma V_m}{r^* R_g T} \tag{5.2}$$

Here, C* is the critical supersaturation, (C–C$_s$), and the other variables were discussed above. Examples of the calculation of r/r* and supersaturation, S, will be shown in the experimental section.

REACTION VARIABLES T, R, C$_s$

In Equation (5.1), reactant addition rate, R, temperature, T (K), and crystal solubility, C$_s$, are the only reaction variables. Of these, the reaction temperature, T, and its measurement control do not need further discussion since they are standard laboratory procedures.

Reactant Addition Rate, R

The reactant addition rate, R (mole/s), is given by two independent but interrelated factors, the reactant concentration, C (mole/cm³), and flow rate, F (ml/s), in Equation (5.3). It is important to note that the R is dependent on both C and F and not on reactant concentration and flow rate alone. We will show examples where the same reaction addition rate was obtained by varying concentration and flow rate, and the resulting crystal number is the same.

$$R = C*F \qquad (5.3)$$

Solubility

Of the three experimental reaction control parameters, maintaining constant solubility provides the greatest challenge. The measurement may differ for varying crystallizations. Electric conductivity or measurement of an electrode potential versus a reference electrode are some useful technologies for ionic crystals. We will discuss the solubility challenge using silver chloride in aqueous solution as an example.

The solubility of ionic crystals is given by the sum of the concentrations of all soluble species in solution that contain the reaction-determining ion, and which participate in the growth of the crystals. In the case of silver chloride, the precipitation is generally performed with chloride excess, so that the silver ion is the reaction-determining ion. Under these reaction conditions, the concentration of growth-determining silver ion, the solubility of the AgCl crystals, C$_s$, is given by the concentrations of free silver ion of silver chloride molecules in solution, and of silver-chloride (AgCl$_m^{1-m}$) complexes in solution as shown in Equation (5.4).

$$C_s = [Ag^+] + [AgCl] + [AgCl_2^-] + [AgCl_3^{2-}] + \left[AgCl_4^{3-} \right] \qquad (5.4)$$

The necessary constants to calculate the concentrations of these species are available from the literature and are compiled in Table 5.1.[1,2] The constants for silver bromide, iodide, and other silver complexes can be found in the cited literature.

In the present experiments, the free silver ion concentration, [Ag$^+$], is determined by the electrochemical potential (EMF, electromotive force) of a silver electrode in the reaction mixture versus a reference Ag/AgCl/4.0 N (mole/l) KCl electrode. The reference electrode is preferentially kept at room temperature (20°C) and connected to the reaction mixture via a liquid junction. Under these conditions, the free silver ion

TABLE 5.1

Constants for AgCl Solubility Calculation

Component	$K_{m,n}$	ΔH
	(25C)	(kcal/mole)
$[Ag^+]$	$10^{K_{sp}}$	See below[a]
$[AgCl]$	$10^{-3.3}$	-2.7
$[AgCl_2^-]$	$10^{-5.25}$	-3.9
$\left[AgCl_3^{2-}\right]$	$10^{-5.7}$	-5.8
$\left[AgCl_4^{3-}\right]$	$10^{-5.4}$	-13.9

$$K_{sp} = [Ag^+][Cl^-] \qquad K_{m,n} = \frac{[Ag^+]^m[Cl^-]^n}{\left[Ag_m Cl_n^{m-n}\right]}; m = 1$$

[a] $\log K_{sp} = -(3206 \pm 42)/T(K) + (1.17 \pm 0.14)$

Errors are one standard deviation; $T(K) = 273.15°C$

concentration, $[Ag^+]$, is given by the electromotive force (vAg in mV) and the temperature (T, °C) of the reaction (Equation [5.5]).

$$pAg = \frac{(599 - vAg) - 0.129(T - 25.0)}{0.198(273 + T)} \qquad (5.5)$$

$$pAg = -\log[Ag^+], T = °C$$

With the knowledge of the solubility product and the free silver ion concentration, the halide concentration can be calculated. Standard physical calculations are then applied to calculate the concentration of the various silver halide complexes from the equilibrium constants and the Gibbs free energy, ΔH (Table 5.1).

In practice, the desired temperature and solubility are chosen for the experiment. The aim EMF is taken from the calculations, and the reaction mixture is adjusted to the aim EMF with a solution of the potassium or sodium salt of the halide. The nucleation EMF is held constant for the estimated nucleation time. After a stable silver halide crystal population is established, the EMF may be changed to modify the morphology of the crystals.

Physicochemical Constants

In the equations, physicochemical constants were presented that can be experimentally determined. For experimental testing of the theoretical models, silver halide precipitations were used. The reason is that the author worked for the photographic industry, where silver halides are the basis for the manufacture of photographic

TABLE 5.2
Physicochemical Constants

Constant	Value	Comment
R_g	8.314E07	Universal gas constant (erg/mole/K)
k_v	1.0 (cube), 0.4714 (octahedra) 4π/3 (sphere)	Volume constant Cube/octahedra: edge length, Sphere: radius
g	140	Surface energy (erg/cm²)
D	(0.88–2.34)E-05	Diffusion constant (cm²/s), 35–80°C
V_m	28.8 (AgCl), 29.0 (AgBr)	Molar volume (cm³/mole)

films. The examples shown for the silver halide precipitations shall be a guide for the evaluation of precipitations of other materials.

The constants used for the calculations are listed in Table 5.2.

- The molar universal gas constant, R_g, is a tabulated standard constant.[3]
- The volume constant, k_v, multiplied with the cube of the related characteristic length, gives the volume of the crystal. The value of k_v changes with the definition of the characteristic size.
- The surface energy, γ, is that of the flat surface area. The surface energy has characteristics for each different type of surface, e.g., 100 (cubic) or 111 (octahedral). In the present case, the experimentally determined surface area for silver halides varies from about 40 to about 160. Thus, the value of 140 was used for the calculations and represents an arbitrary choice.
- The diffusion coefficient, D, should ideally be determined for each growth unit, e.g., the various silver chloride complexes. However, such data were not available. In agreement with the suggestion of Strong and Wey (1979), the diffusion constant of water solvent was chosen for the calculations. The diffusion constant is a function also of the temperature, and a linear correlation with temperature was used.[4]
- The molar volume, V_m, is the volume per mole of crystal material. It is obtained by multiplying the density (cm³/g) with the molecular (or formula) weight.

REACTION VOLUME

The reaction volume is not a factor or variable in Equation (5.1), suggesting that the result of a crystallization should be independent of the reaction volume if temperature, addition rate, and solubility are held constant.

This is in contradiction to the classical nucleation model, where the nucleation rate, J, is normalized to the reaction volume. In our first publication, the addition rate and crystal number were erroneously normalized to the reactor volume.[5] This was corrected in a subsequent publication.[6]

TABLE 5.3
Crystal Size vs. Initial Reaction Volume

	Initial Reaction Volume (L)	Size (μm)	Std. dev. (μm)	# Exp.
AgBr	1.5	0.128	0.005	5
	3.0	0.128	0.003	7
	4.5	0.124	n. a.	1
AgCl	0.75	0.197	n. a.	1
	1.50	0.195	n. a.	1

Variables: reaction volume, reaction addition rate; others: not varied; stirring adjusted

To test the prediction of Equation (5.1), that the crystal nucleation is independent of reaction volume, several experiments were performed in which the initial reaction volume was varied, but Temperature, T, and solubility, C_s, were not varied. The reaction addition rate was varied, but was held constant throughout the experiments. Over time, gelatin batches had to be changed, and any variations due to changing gelatin batches are reflected in the standard errors. For larger reaction volumes, mixing/stirring was increased to avoid mixing artifacts. The results are compiled in Table 5.3 for precipitations of AgCl and AgBr. Several of these experiments were repeated at irregular time intervals, spanning several years. Further, reaction vessels of different shapes were used, such as flat-bottomed or fluted containers.

The data in Table 5.3 support the prediction of Equation (5.1) that the crystal size is independent of the initial reaction volume. Where repeat experiments were performed, standard deviations of the sizes could be obtained. These include variability in preparation and size measurement. The results are clearly independent of initial reaction volume within the standard deviations. Moreover, it also shows that reactor shape did not affect the outcome of the precipitations.

Constancy of Crystal Number

The BNG model predicts that under diffusion-controlled growth conditions, formation of new crystals ends after the nucleation period. This prediction was checked for precipitations of AgCl, AgBr (cubic and octahedral), and AgI. The reaction conditions are included in Table 5.4. Samples were taken during the precipitation. The crystal sizes were determined, and the crystal numbers were calculated from the mass balance.[7] The results are shown in Table 5.4 and in Figure 5.1.

The plot of the crystal numbers versus the time into the precipitation shows that the number of crystals was stable after about 2 minutes. For the first sample, which was taken at 1 minute, the crystal number is larger than the average for AgCl and AgBr. This may be an indication of a transition condition from nucleation to growth, or due to errors to determine the very small crystal sizes.

TABLE 5.4

Crystal Number and Size vs. Time and Reaction Conditions

Time (min)	AgCl Size (μm)	AgCl Cryst. no. (Z × E-15)	AgBr (cubic) Size (μm)	AgBr (cubic) Cryst. no. (Z × E-15)	AgBr (oct) Size (μm)	AgBr (oct) Cryst. no. (Z × E-15)	AgI Size (μm)	AgI Cryst. no. (Z × E-15)
1.0	0.063	4.14	0.044	13.60	0.058	12.60	n.a.	n.a.
2.0	0.104	1.84	0.070	6.76	0.091	6.53	n.a.	n.a.
4.0	0.131	1.84	0.084	7.82	0.109	7.60	0.035	77.3
6.0	0.141	2.21	0.094	8.37	0.118	8.99	0.037	98.1
8.0	0.153	2.31	0.105	8.01	0.129	9.17	0.044	77.7
12.0	0.173	2.40	0.119	8.26	0.146	9.48	0.047	95.7
16.0	0.196	2.20	0.126	9.27	0.153	11.00	0.059	64.5
20.0	0.215	2.08	0.137	9.02	0.167	10.60	0.063	66.2
Average		2.13		8.22		9.05		79.9
Std. dev.		0.22		0.83		1.57		14.3
pAg (70°C)	5.85		6.95		8.75		9.50	

Initial reactor volume: 2.0 L; 2.5% deionized gelatin; 70°C; pAg (see above); reactant addition rate: 0.040 mole/min, AgI: 0.020 mole/min

The predictions and experimental results of the BNG model have provided important information for practical crystallization experiments:

1. Crystal size and number are independent of reaction volume.
2. After a short nucleation time, the crystal number is constant.
3. The crystal number is only dependent on the reaction addition rate. The crystal size increases with the length of reaction addition.
4. Flow rate and input reaction addition rate individually are not nucleation controlling.

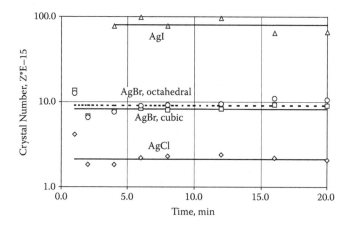

FIGURE 5.1   Crystal number versus precipitation time.

MODELING: SEPARATION OF VARIABLES R, C$_S$, AND T

THEORY

In this and the next sections, the change of crystal size and number will be experimentally related to the reaction control variables addition rate, R, crystal solubility, C$_s$, and temperature, T. For this purpose, Equation (5.6) needs to be separated into the individual control parameters.

Let's return to Equation (5.6), the correlation between number of stable crystals formed, Z, and reaction variables addition rate, R, Temperature, T, and solubility, C$_s$. The diffusion constant, D, is temperature dependent and must be included with the temperature as a control variable.

$$Z = \left[\frac{R_g}{2k_v \gamma D V_m}\right] \frac{RT}{C_s(r/r^* - 1.0)} \tag{5.6}$$

Taking the logarithm of Equation (5.6) leads to Equation (5.7), where the control variables are separated.

$$\ln Z = \ln R - \ln C_s + \{\ln T - \ln \gamma - \ln D\} - \ln(r/r^* - 1.0) + \ln\left[\frac{R_g}{2k_v V_m}\right] \tag{5.7}$$

In these equations, r/r* may depend on the addition rate, temperature, and solubility. The diffusion coefficient, D, and the surface energy, γ, are also temperature dependent, and these effects must be considered with the direct temperature effect, T, as indicated in Equation (5.7). In the next sections, the temperature dependence of D and γ will be reviewed before proceeding to the analytical derivations.

Diffusion Coefficient, D

The Einstein model for diffusion of a particle in a solvent is given by Equations (5.8) and (5.9).

$$D = \frac{R_g T}{6\pi N \eta r} \tag{5.8}$$

$$\ln D = \ln T - \ln \eta + \ln\left(\frac{R_g}{6\pi N r}\right) \tag{5.9}$$

N is the Avogadro constant, R$_g$ the gas constant, T the temperature (K), η the viscosity of the medium (solvent), and r the radius of a spherical particle that is large compared to the solvent molecules.

The viscosity of the medium, η, decreases with temperature. For water, its temperature dependence between 10° and 100°C can be well described by Equation (5.10).[8]

$$\ln\eta = \frac{1781}{T} - 12.97 \tag{5.10}$$

The correlation coefficient, R^2, is equal to 0.9950.

Surface Energy, γ

Surface energy, γ, is a function of temperature. For that reason, when a value is given for the surface tension of an interface, temperature must be explicitly stated. The general trend is that surface tension decreases with an increase of temperature, reaching a value of zero at a critical temperature, and that the decrease is approximately linear.[9] There are only empirical equations to relate surface tension and temperature, e.g., Equation (5.11).[10,11]

$$\gamma = k(T_C - T)/V_m^{2/3} \tag{5.11}$$

V_m is the molar volume of that substance, T_C is the critical temperature, and k is a constant for each substance. For example, for water k = 1.03 erg/°C (103 nJ/K), V_m = 18 ml/mol, and T_C = 374K (100°C). For solids we will assume that the surface energy does not change significantly between the freezing and boiling temperatures of water.

The stage is now set to separate Equation (5.7) into its components to determine the dependence of the crystal number, Z, on the reaction variables addition rate, R, solubility, C_s, and temperature, T.

ADDITION RATE, R

When Equation (5.7) is differentiated for the addition rate, R, constant temperature, T, and solubility, Equation (5.12) is obtained.

$$\frac{d\ln Z}{d\ln R} = 1.0 - \frac{R}{r/r^* - 1.0} * \frac{d(r/r^*)}{dR} \tag{5.12}$$

Whether the slope is equal or different from 1.0 is determined by the differential $d(r/r^*)/dR$. For experimental conditions where r/r^* is independent of addition rate, a linear increase of crystal number with increasing addition rate is predicted.

For a change of r/r^* with addition rate, the possible effect of addition rate on the supersaturation during nucleation must be considered. An increase of the addition rate may increase the supersaturation, S, which is the difference between the actual concentration, C, and the equilibrium saturation concentration, C_s (Equation [5.13]).

$$R \sim S = C - C_s \tag{5.13}$$

The critical crystal size, r*, is related to the supersaturation through the Gibbs-Thomson equation (Equation [5.14]).

$$r* = \frac{2\gamma V_m C_s}{R_g T(C - C_s)} \sim \frac{1.0}{R} \tag{5.14}$$

Returning to Equation (5.12), it follows that the ratio r/r* may increase with R, but not decrease. Thus, it can be concluded that a plot of ln Z versus ln R (or equivalent log/log plots) should yield a linear correlation with a slope equal or less, but not higher than 1.0.

In the experimental section, we will show examples in which precipitations of AgCl and AgBr yield a linear slope. However, it has also been reported that for AgBr precipitations under different conditions, slopes of 0.776, 0.942, and 0.993 have been obtained.[12] Slopes greater than 1.0 have not been reported.

SOLUBILITY, C_s

Differentiating Equation (5.7). for $\ln C_s$ yields Equation (5.15).

$$\frac{d\ln Z}{d\ln C_s} = -\left(1.0 + \frac{C_s}{r/r* - 1.0} * \frac{d(r/r*)}{dC_s}\right) \tag{5.15}$$

If $d(r/r*)/d\,C_s$ equals zero, the plot of log Z versus log C_s will be linear and have a slope of minus one. Experimental plots for silver chloride show this type of correlation.

$$r* = \frac{2\gamma V_m}{R_g T(C/C_s - 1.0)} \sim C_s \tag{5.16}$$

If r* increases with C_s, the second part in Equation (5.15) becomes negative, and the value in the large brackets becomes less than one. Thus, it is predicted that a plot of log Z versus log C_s will have a slope of −1.0 or more positive. Silver bromide will provide an example of this condition and predictions.

TEMPERATURE, T

Differentiating Equation (5.7). for ln T yields Equation (5.17).

$$\frac{d\ln Z}{d\ln T} = 1.0 - \frac{T}{r/r* - 1.0} * \frac{d(r/r*)}{dT} - \frac{d\ln\gamma}{d\ln T} - \frac{d\ln D}{d\ln T} \tag{5.17}$$

The factor $d(r/r*)/dT$ increases with temperature, since r* decreases with increasing temperature (Equation [5.16]). For solids, it is assumed that the surface energy does not change significantly for the temperature range of water, and $d\ln\gamma/d\ln T$ may approximately set equal to zero. The factor $d\ln D/d\ln T$ increases with temperature, and is set equal to 1.0, since the change in water viscosity probably does not significantly change the temperature dependence of D (Equations [5.9] and [5.10]). This gives a correlation of the $d\ln Z/d\ln T$ correlation as Equation (5.18).

$$\frac{d\ln Z}{d\ln T} = -\frac{T}{r/r^* - 1.0} * \frac{d(r/r^*)}{dT} \qquad (5.18)$$

Equation (5.18) indicates that the number of crystals formed decreases with increasing temperature. This is in agreement with experimental results for silver chloride and bromide, and is generally observed.

ADDITION RATE

It is the aim of this section to test the predictions of the BNG theory with regard to the correlation between number of stable crystals formed and reaction addition rate under controlled precipitation conditions.

To separate the addition rate from the other control variables, Equations (5.6)–(5.12) show that the temperature and solubility must be held constant for the experiments. Equation (5.20) shows how to calculate the unknown, r/r^*, from the experimental results. Preferentially, the inverse, r^*/r, is tabulated and discussed, since its limiting values are zero and 1.0. A value of zero indicates continuous nucleation, while values equal to or greater than 1 are associated with a breakdown or significant change of the nucleation process. An example of the latter will be discussed later.

$$Z = \left[\frac{R_g}{2k_v \gamma D V_m}\right] \frac{RT}{C_s (r/r^* - 1.0)} \qquad (5.19)$$

$$\frac{r}{r^*} = \left[\frac{R_g}{2k_v \gamma D V_m}\right] \frac{RT}{Z C_s} + 1.0 \qquad (5.20)$$

Analysis of Equation (5.6) leads to Equation (5.21) where it was determined that the slope of the log Z versus log R correlation would yield a slope of one or less.

$$\frac{d\ln Z}{d\ln R} = 1.0 - \frac{R}{r/r^* - 1.0} * \frac{d(r/r^*)}{dR} \qquad (5.21)$$

To test the predictions, silver chloride, AgCl, and silver bromide, AgBr, were precipitated under conditions where the solubility, as measured by the pAg, and temperature were held constant.[5,6,13]

Since the experiments stretched over many years, the precipitations occurred in reaction vessels of different sizes and shapes and with different initial reaction volumes. It was shown previously that a change in initial reaction volume did not affect the outcome of the crystallization process. The erroneous normalization of the crystal number to the initial reaction volume[5] was corrected in following publications.[7]

TABLE 5.5
AgCl Variation of Addition Rate and Results

No.	Addition rate (mmole/min)	Flow Rate (ml/min)	Concentr. (mole/l)	Size (μm)	Cryst. no. (Z × E14)	r*/r
1	2.5	10	0.25	0.190	1.8	0.55
2	10.0	40	0.25	0.200	6.5	0.53
3	10.0	10	1.00	0.190	7.6	0.56
4	20.0	20	1.00	0.200	13.1	0.53
5	40.0	40	1.00	0.190	32.0	0.58
6	40.0	10	4.00	0.200	28.0	0.54
7	80.0	20	4.00	0.200	54.0	0.54
8	160.0	40	4.00	0.200	99.0	0.51
9	240.0	60	4.00	0.200	146.0	0.51
10	320.0	80	4.00	0.200	195.0	0.51
			Average	0.197		0.54
			Std. dev.	0.005		0.02

Also, certain addition rates were obtained by simultaneously changing reactant concentration and flow rate. This information is contained in Table 5.5 for silver chloride and in Table 5.6 for silver bromide. By changing the addition rate over several orders of magnitude, high accuracy and reliability in the determination of the log Z/log R correlation was assured.

TABLE 5.6
AgBr, Variation of Addition Rate and Results

No.	Addition Rate (mmole/min)	Reaction Volume (liter)	Concentr. (mole/l)	Size (μm)	Cryst. no. (Z × E16)	r*/r
1	0.2	3.0	0.010	0.129	0.0054	0.53
2	1.0	3.0	0.050	0.133	0.0244	0.51
4	2.0	3.0	0.100	0.127	0.0570	0.55
5	8.0	3.0	0.400	0.128	0.2188	0.54
6	16.0	1.5	0.800	0.130	0.4229	0.53
7	16.0	3.0	0.800	0.125	0.4760	0.56
8	20.0	3.0	1.000	0.126	0.5744	0.55
9	32.0	1.5	1.600	0.121	1.0477	0.58
10	32.0	3.0	1.600	0.125	0.9500	0.61
11	32.0	4.5	1.600	0.124	0.9800	0.60
12	64.0	1.5	3.200	0.126	1.8557	0.55
13	96.0	1.5	4.800	0.131	2.4768	0.52
14	128.0	1.5	6.400	0.132	3.2486	0.52
			Average	0.127		0.55
			Std. dev.	0.003		0.03

SILVER CHLORIDE

Experimental: AgCl

Precipitation Conditions

The initial reaction volume was 1.20 liter distilled water. A relatively high concentration of 4% deionized bone gel was dissolved to assure that sufficient peptization was provided for the silver halide formed at the highest reaction addition. The reaction mixture was adjusted to 60°C and pAg 6.4, which corresponds to a solubility of $5.4 \times E-06$ mole Ag^+/l. The concentration and flow rate of the reactants are listed in Table 5.5. The reactant addition time was 20 minutes. After the end of precipitation, an aqueous solution of the sodium salt of TAI (4-hydroxy-6-methyl-1,3,3a,7-tetraazaindine) was added to restrain growth by Ostwald (= physical) ripening. The crystal sizes were determined from electronic imaging analysis and Joyce Loebl Disk Centrifuge.[14] For the determination of the crystal number, Z, the different sizing methods were averaged.

Stirring/Mixing

Special consideration for controlled crystallizations must be given to effective mixing of the reactants with the reaction medium and with each other. Another concern is effective stirring of the reaction medium to avoid dead zones, and to avoid settling of reactant. In our experiments, agitation was provided by a center-mounted, counter-acting, radial flow, double-shrouded turbine. The mixer consisted of two shrouded turbines separated by a spacer. The two turbines were arranged to direct their flow toward this center space. The two reactant solutions were separately introduced below the surface of the suspension through tubes directed toward the ending at the mixer head. This setup provided optimum mixing of the reactants with the reaction medium and intimate mixing of the reactant streams at the center of the turbine assembly. After mixing, the mixture streamed toward the side of the reactor and provided efficient stirring, i.e., reactor turnover. Stirring and mixing were further enhanced by adding baffles into the reactor. The baffles were provided with holes to provide added turbulence, and thus increased mixing of the reaction mixture. In addition, the baffles reduced the formation of a vortex in the center of the reactor. The stirring rate of the turbine was slowly increased until a vortex formed, and then reduced just below vortex formation. As the reaction volume increased due to reactant addition, this procedure was repeated to continue to provide maximum stirring/mixing conditions for the system.

It was experimentally determined that addition to the surface or subsurface of the reaction mixture provided poor mixing and led to significant variability of crystal sizes and size distribution.

Peptizer

In agitated systems, interaction between particles may lead to undesired agglomeration. For systems like the present ones, in which ions of opposite charge are added, the surface of the crystals becomes charged. Thus, in the case of $AgNO_3$ addition, crystals will be positively charged, and for the halide addition, crystals will be negatively charged. Their interaction will lead to aggregation and agglomeration. Under further growth, agglomerated/aggregated crystals will grow together at the contact site. These processes will lead to uncontrolled morphology of the resultant crystal population.

To prevent agglomeration, materials are added that adsorb to the crystal surface to reduce direct interaction and to provide steric hindrance for the interaction of the surfaces of neighboring crystals. In the case of silver halides, gelatin, an organic macromolecule, has been known to show surface protection and other desirable properties. For other materials, their adsorption characteristics and interactions with large protective molecules must be determined individually.

Calculation of r*/r

For the calculation of r*/r, the following values were used: gas constant, R_g, 8.3 × 10^7 erg/°C × mole; molar volume, V_m, 25.9 and 29.0 for AgCl and AgBr, respectively; surface energy, γ, 140 erg/cm^2; for the diffusion coefficient, D, a linear increase from 0.88 to 2.34 × 10^{-5} between 35° and 80°C was used as suggested by Strong and Wey.[4] The value of the diffusion coefficient is that of water, since the diffusion coefficients of halide and silver ions and their temperature dependence have not yet been determined.

Results

Table 5.5 summarizes the experimental results for changes in crystal number and size of silver chloride as a function of addition rate. The addition rate varied from 2.5 to 320 mmole/min, which is a change of approximately two orders of magnitude. This wide experimental range gives high confidence to test the validity of the predictions of the BNG nucleation model.

The average crystal size is 0.197 μm with a standard deviation of 0.005 μm. For the ratio of r*/r, an average value of 0.54 +/−0.02 is obtained. This indicates that the critical crystal size is about 54% of the average crystal size throughout the precipitation after the nucleation phase. The consistency of these results shows high reproducibility of the experimental conditions.

Experiments 2 and 3 and 5 and 6 are replicas in which the reactant concentration and flow rate were varied while maintaining the reactant addition rate. The results of 0.20 and 0.19 and 0.19 and 0.20 μm for size variability and 0.53 and 0.56 and 0.58 and 0.54 for r*/r are identical well within the ranges of error. This consistency supports the claim of the theory that the reactant addition rate, not the flow rate or reactant concentration, controls the nucleation process.

Figures 5.2 and 5.3 show the correlations of experimental results as a function of addition rate. Figure 5.2 shows that the number/addition rate correlation is linear. This linear correlation indicates that the ratio r*/r is a constant (as also shown by the discussion above) and that r* is not a function of the addition rate. Figure 5.2 also shows the plot of crystal size and of r*/r versus the addition rate and illustrates that these values are constant over the full range of addition rates.

Figure 5.3 shows a log Z/log R plot of the data, which complements the linear plot of Figure 5.2. The latter compresses many of the data in the lower addition rate range and may hide nonlinearity. The log/log plot of Figure 5.3 gives more equal weight to the whole ranges of crystal number and addition rate. The linearity of the

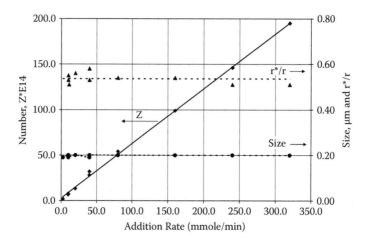

FIGURE 5.2   Silver chloride crystal number, size, and r*/r versus addition rate.

correlations is supported by statistical correlation of the data in Equations (5.22) and (5.23).

$$Z = (0.607 \pm 0.014)10^{17}R + (1.81 \pm 1.66)10^{14}$$

$$R^2 = 0.999 \tag{5.22}$$

$$\log Z = (0.960 \pm 0.030)\log R + (16.78 \pm 0.14)$$

$$R^2 = 0.999 \tag{5.23}$$

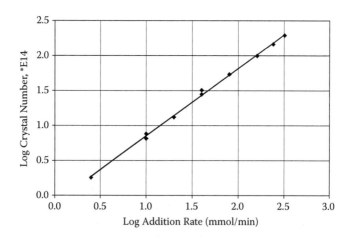

FIGURE 5.3   Silver chloride log crystal number versus log addition rate.

Here, the limits of error represent one standard deviation, which will be retained throughout the chapter. Z is the crystal number and R is given in mole/min.

SILVER BROMIDE

The data for silver bromide were obtained over several years and thus reflect the changes in precipitation technology. The control of addition rate, temperature, and solubility (pAg) during the experiments improved due to the availability of electronic control versus initial manual control by the experimenter. Thus, the initial data indicated a slope less than 1.0 for the log Z/log R correlation, which was in error due to difficult control of low addition rates. Such problems were overcome in the later experimental setups.

Experimental: AgBr

The experimental setup was the same for AgBr precipitations as described for the AgCl precipitations. For the silver bromide precipitations, initial reaction volumes of 1.5, 3.0, and 4.5 liters were used. As predicted by the BNG model, these differences did not significantly affect the number of crystals or final crystal sizes. The initial reaction mixture was composed of an aqueous solution of 4% deionized bone gelatin. The pAg was adjusted with sodium bromide at the aim temperature. The addition rate was held constant for the 20-minute reaction time. Differently from the AgCl precipitations, the addition flow rate was held constant at 20 ml/min.

Results

The experimental variations and results are summarized in Table 5.6. The results show that the crystal size did not significantly vary from the average of 0.197 μm between the addition rates of 2.5 to 320 mmole/min. Experiments 5 and 7 and 9 and 10 are repeats in which the initial reaction volume was varied between 1.5 and 3.0 liter. The resulting crystal sizes and numbers are the same within the error of limits. Thus, these results confirm that reaction volume is not a determining factor for crystal size and number formed during nucleation.

Throughout the same set of experiments, the ratio of r*/r (0.54) indicates that the critical crystal size is about 54% of the average crystal size and does not significantly vary.

Figure 5.4 shows a linear plot of crystal number, size, and r*/r versus addition rate. The plot visually confirms that size and r*/r are independent from addition rate. The number of crystals increases linearly as predicted by theory with the variation of addition rate over two orders of magnitude.

The log Z/log R plot of Figure 5.5 expands the low-addition rate data versus the linear plot of Figure 5.4. It confirms the theoretical prediction of a linear log Z versus log R correlation.

Quantitative evaluation of the results gives the correlations of Equations (5.24) and (5.25).

$$Z = (25.7 \pm 0.7)10^{13}R + (51.9 \pm 92.3)10^{13} \tag{5.24}$$

$$\log Z = (1.01 \pm 0.01)\log R + (14.432 \pm 0.089) \tag{5.25}$$

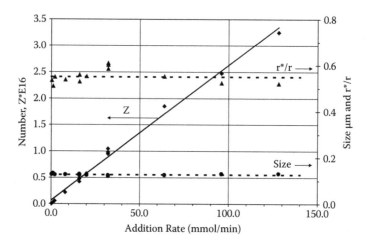

FIGURE 5.4   Silver bromide crystal number, size, and r*/r versus addition rate.

The intercept of the Z/R correlation is equal to zero, and the slope of the log Z/log R correlation is equal to 1.0, which are within the limits of error. The intercept of the log Z/log R correlation predicts the number of crystals for an addition rate of 1.0.

These experimental results are also in agreement with the prediction of the basic derivation of the BNG nucleation phase (Chapter 3).

Practical Applications

Practical application of these experiments predict that scaling of a precipitation with the aim of obtaining reaction-independent crystal sizes is optimally achieved when temperature, crystal solubility, and optimum mixing/stirring are maintained while varying the reactant addition rate and time. Under these conditions, it is shown that

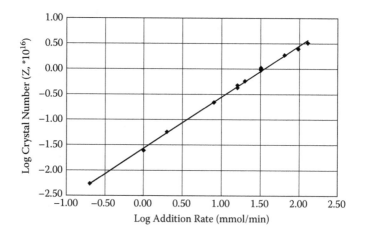

FIGURE 5.5   Silver bromide log crystal number versus log addition rate.

initial reaction volume does not change the crystal size. Included in the presented reaction schemes is the observation that geometric changes of the reaction vessel, e.g., from cone to flat-bottom, did not change the results of the precipitations.

For experiments in which the slope of the log Z/log R correlation is less than one, the same rules apply. However, under these conditions, which are otherwise identical, the crystal size will increase with addition rate while crystal morphology is maintained.

In conclusion, the predictions of the BNG model with respect to the dependence of the crystal number and size on the reactant addition rate during nucleation have been experimentally confirmed.

SOLUBILITY

Varying the solubility of the crystals is a simple and effective procedure to vary the crystal size and often also the morphology of the precipitate. The solubility is explicitly included in the fundamental derivation of the BNG model (Equation [5.6]). Separating the variables and differentiating the crystal number as a function of solubility lead to Equation (5.15).

This equation predicts that the number of crystals nucleated, Z, decreases with increasing solubility. If $r/r*$ is independent of C_s, then the differential $d(r/r*)/dC_s$ is equal to zero, and Z is predicted to decrease linearly with solubility, with a log Z/log C_s slope of -1.0.

However, if $r/r*$ is dependent on C_s, Z is nonlinearly correlated with C_s. In this case, the theory predicts that the log Z/log C_s correlation has a slope different from -1.0. Theoretical predictions indicate that under such conditions, $r*$ should increase with solubility, predicting a log Z/log C_s more positive than -1.0.

$$Z = \left[\frac{R_g}{2k_v \gamma DV_m} \right] \frac{RT}{C_s (r/r*-1.0)} \tag{5.26}$$

$$\frac{d\ln Z}{d\ln C_s} = -\left(1.0 + \frac{C_s}{r/r*-1.0} * \frac{d(r/r*)}{dC_s} \right) \tag{5.27}$$

We will explore these predictions for controlled precipitations of silver chloride and silver bromide.

SILVER CHLORIDE

Experimental

The dependence of crystal nucleation on solubility was adjusted by varying the pAg of the reaction mixture at aim temperature with a solution of sodium chloride. The solubility of silver halides is a function of pAg and temperature. The reaction temperature was 60°C, and for the adjustment of solubility the pAg was varied from 4.95 to 8.30. In this range, the solubility has a minimum around pAg 6.4; further details

TABLE 5.7
Silver Chloride Nucleation Dependence on Solubility

No.	Solubility (mole/l)	pAg	Size (μm)	Number (Z)	r*/r
1	1.50E-05	4.95	0.271	2.60E+14	0.55
2	1.10E-05	5.10	0.265	2.80E+14	0.49
3	7.50E-06	5.35	0.205	6.00E+14	0.59
4	6.10E-06	5.60	0.196	6.90E+14	0.57
5	5.00E-06	5.90	0.195	7.00E+14	0.53
6	5.40E-06	6.45	0.195	7.00E+14	0.54
7	6.80E-06	6.75	0.204	6.10E+14	0.57
8	1.00E-05	7.00	0.227	4.40E+14	0.61
9	1.20E-05	7.10	0.254	3.20E+14	0.55
10	1.60E-05	7.30	0.275	2.50E+14	0.56
11	2.90E-05	7.55	0.340	1.30E+14	0.54
12	5.30E-05	7.80	0.416	7.20E+13	0.55
13	2.30E-04	8.30	0.825	9.30E+12	0.40
				Average	0.54
				Std. dev.	+/−0.05

Temperature, 60°C; addition rate, 0.010 mole/min; initial reactor volume, 0.74 l; reactant flow rate, 10 ml/min; reaction time, 20 min; morphology: cubic

were discussed in the previous section and are listed in Table 5.7.[6] The experiment at pAg 8.3, the highest solubility, gave a significantly larger crystal size than indicated by the other data points. This deviation may be due to experimental variability. However, it may also indicate a change in the mechanism of crystal nucleation and growth, as indicated by the change from cubic to octahedral crystal shape for silver bromide. Because of these uncertainties, the data point is reported, but it was not used for the quantitative correlations.

Results: AgCl

The results of the experiments are listed in Table 5.7 and are plotted in Figures 5.6 through 5.10.

Equation (5.6) predicts that the number of crystals formed should be inversely proportional to the silver chloride solubility. Figure 5.6 shows a plot of the crystal number and of the solubility as a function of pAg. It is apparent that the crystal number has its maximum around pAg 6.4, where the solubility has its minimum. As the solubility increases from its minimum at lower and higher pAg, the number of crystals formed decreases.

Conversely, since the crystal number is proportional to 1/size,[3] the crystal size is predicted to be proportional to the solubility. This is evident from Figure 5.7, where crystal size and solubility are plotted as a function of pAg, and where the crystal

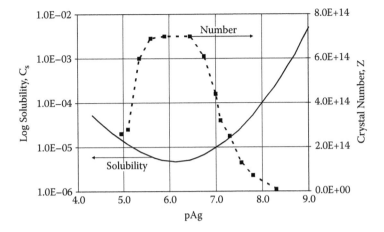

FIGURE 5.6 Silver chloride crystal number and solubility as a function of pAg.

number minimum coincides with minimum solubility and increases with increasing solubility. Figures 5.6 and 5.7 thus confirm the prediction of Equation (5.6).

Equation (5.15) predicts that the logarithm of the crystal number should decrease linearly with the logarithm of solubility. The plot in Figure 5.8 shows that a linear correlation is obtained as predicted. For silver chloride, the data fall on the same line independent if the solubility was below or above minimum solubility. We will see that this is not the case for silver bromide.

Equation (5.6) and the results of Figure 5.8 suggested that a linear correlation may exist between crystal number and inverse solubility, and this correlation is shown in Figure 5.9. The trend line gives a linear correlation within the limits of error.

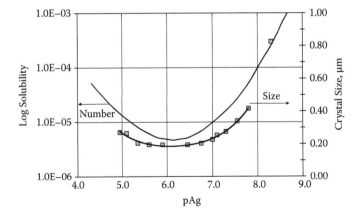

FIGURE 5.7 Silver chloride crystal size and solubility as a function of pAg.

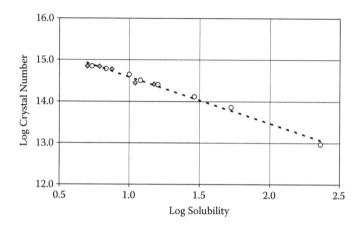

FIGURE 5.8   Silver chloride log crystal number as a function of log solubility.

The correlations between crystal number and solubility are quantified in Equations (5.28) and (5.29).

$$logZ = (-1.126 \pm 0.038)*logC_s + (8.957 \pm 0.186)$$

$$R^2 = 0.9937 \tag{5.28}$$

$$Z = (3.888 \pm 0.220)10^9*(1/C_s) + (0.33 \pm 2.56)10^{13}$$

$$R^2 = 0.9828 \tag{5.29}$$

Both equations and the correlation coefficients show that the predictions of Equation (5.6) and (5.15) are confirmed by the experimental results.

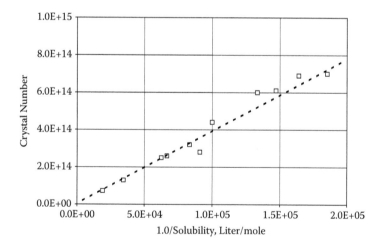

FIGURE 5.9   Silver chloride number as a function of 1.0/solubility.

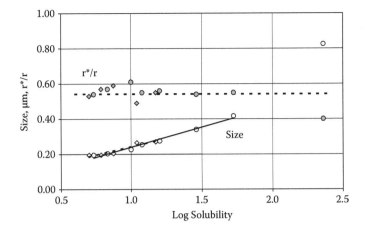

FIGURE 5.10 Silver chloride crystal size and r*/r as a function of log solubility.

Crystal size and r*/r are plotted versus log solubility in Figure 5.10. The plot shows that r*/r is a constant and independent of solubility for the given reaction conditions with an average of 0.54 and a standard deviation of 0.05.

Under the same conditions, the size increases linearly with log solubility, as confirmed by Equation (5.30) and the correlation coefficient.

$$\log Size = (0.3756 \pm 0.0130)\log C_s + (1.2534 \pm 0.0636)$$

$$R^2 = 0.9934 \tag{5.30}$$

The experimental results for precipitations of silver chloride at a wide range of solubility thus confirm the predictions of the balanced nucleation and growth model.

SILVER BROMIDE

The situation for silver bromide is more complicated than that for silver chloride. The morphology of the crystals changes between cubic and octahedral and is related to the minimum solubility. Thus, the BNG model was studied for both morphologies.

Experimental

To determine the dependence of crystal number, crystal size, morphology, and r*/r as a function of solubility, the temperature of 70°C was chosen. The pAg and solubility were adjusted at aim temperature with a solution of sodium bromide. The gelatin concentration was 4% for the initial reaction volume of 1.5 liter. The temperature and the addition rate of 0.032 mole/min were held constant for the reaction time of 20 minutes. Samples were taken at the end of the precipitation, and physical ripening of the crystals was minimized by the addition of 1-phenyl-5-mercaptotetrazole, a growth restrainer for silver halide.[5,6]

TABLE 5.8

Silver Bromide Nucleation Dependence on Solubility

AgBr No.	pAg	Solubility (mole/l)	Size (µm)	Number (Z)	r*/r	Morphology
1	6.30	3.20E-06	0.165	4.13E+15	0.64	cubic
2	6.70	1.60E-06	0.138	7.08E+15	0.60	cubic
3	7.10	1.00E-06	0.123	9.95E+15	0.57	cubic
4	7.60	8.60E-07	0.118	1.13E+16	0.57	cubic
5	8.00	1.00E-06	0.135	1.61E+16	0.68	octahedral
6	8.40	1.60E-06	0.138	1.58E+16	0.77	octahedral
7	8.60	2.10E-06	0.146	1.28E+16	0.78	octahedral
8	8.80	3.10E-06	0.161	9.62E+15	0.80	octahedral
9	9.25	1.00E-05	0.178	6.96E+15	0.90	octahedral

Temperature, 70°C; addition rate, 0.032 mole/min; initial reactor volume, 1.5 l; reaction time, 20 min

Results: AgBr

The experimental results are listed in Table 5.8 and Figures 5.11 to 5.14. In addition to the reaction variables and results, the table also contains the information on the morphology of the crystals.

In Figure 5.11, the number of crystals and the solubility are plotted versus pAg. The solubility shows the minimum of solubility at about pAg 7.3. Cubic crystals are obtained up to pAg 7.6. Between pAg 7.6 and 8.0, the transition from cubic to octahedral morphology occurs, accompanied by a major increase of crystal number. This transition signals a major shift in the growth and nucleation mechanism.

The crystal number mirrors the change of solubility, as predicted by Equation (5.6). A significant difference in comparison to silver chloride is the shift of the

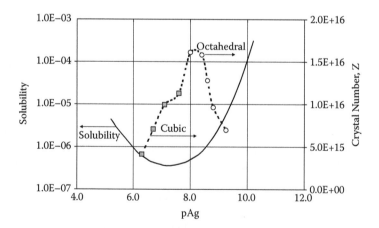

FIGURE 5.11   Silver bromide crystal number and solubility as a function of pAg.

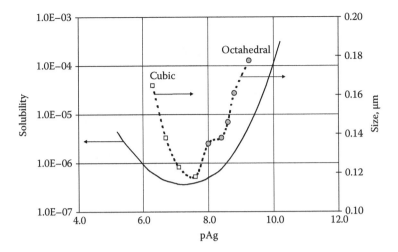

FIGURE 5.12 Silver bromide crystal size and solubility as a function of pAg.

maximum of crystal number (pAg 8.2) to higher pAg than the minimum of solubility (pAg 7.6). This shift is probably associated with the change in morphology.

Figure 5.12 shows the variation of crystal size with pAg and solubility. The crystal size follows the change in solubility, as already discussed for silver chloride. However, unlike silver chloride, silver bromide has a size-solubility correlation that appears less tight.

In Figure 5.13, the validity of Equation (5.15) is tested for silver bromide. The plot of log crystal number (Z) versus log solubility (C_s) shows two distinctly different, but linear, correlations for the cubic and the octahedral crystals. This indicates that r*/r may be dependent on solubility and affect the slope of the correlation. Equations (5.31) and (5.32) give the correlations between log Z and log C_s. The significantly

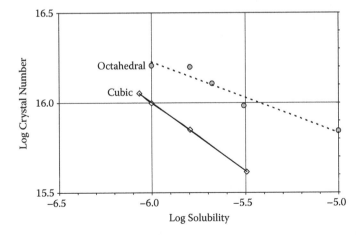

FIGURE 5.13 Silver bromide log crystal number as a function of log solubility.

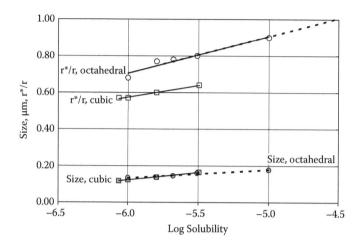

FIGURE 5.14   Silver bromide crystal size and r*/r as a function of log solubility.

lower slope than the ideal −1.0 indicates that for both cubic and octahedral crystals r*/r has a significant effect.

$$\log Z(\text{cub}) = -(0.761 \pm 0.009)\log C_s + (11.435 \pm 0.053)$$

$$R^2 = 0.9998$$

(5.31)

$$\log Z(\text{oct}) = -(0.397 \pm 0.058)\log C_s + (13.848 \pm 0.325)$$

$$R^2 = 0.9693$$

(5.32)

Thus, crystal size and r*/r are plotted as a function of log solubility in Figure 5.14. The crystal sizes vary relatively little over the solubility range. However, the correlation for r*/r shows significant slope.

The correlations of r*/r versus log solubility are given in Equations (5.33) and (5.34) for cubic and octahedral crystals, respectively.

$$\log(r^*/r)(\text{cub}) = (0.093 \pm 0.007)\log C_s + (0.314 \pm 0.040)$$

$$R^2 = 0.9946$$

(5.33)

$$\log(r^*/r)(\text{oct}) = (0.110 \pm 0.018)\log C_s + (0.511 \pm 0.103)$$

$$R^2 = 0.9609$$

(5.34)

The value of r*/r increases significantly over the solubility range for both morphologies. For cubic crystals, the solubility reaches a maximum at pAg 5.20 (solubility

4.40E–06), and an extrapolation of the r*/r correlation to higher solubilities is limited.

For octahedral crystals, the solubility limit for cubic crystals does not apply. Thus, it is possible to extrapolate to where the value of r*/r reaches 1.0, which is at log solubility of −4.65 and at pAg 9.7, respectively.

Under conditions where r* is equal to or greater than r, and r*/r is equal to or larger than 1.0, the system is unstable. Thus, it is predicted that for a log solubility value more positive than −4.54, which is equal to pAg of 9.7, crystals of octahedral morphology are no longer stable.

AgBr Morphology as a Function of Solubility

In the previous section, we have shown that the BNG model predicts that for silver bromide, the critical crystal size would exceed the limiting r*/r ratio of 1.0 at the solubility of about $10^{-4.5}$ mole/l (log−4.65). This prediction was tested by precipitations increasing the solubility from about 10^{-6} up to 10^{-4} (mole/l; pAg 8.0 to 9.1).

Electron micrographs were obtained for each experiment. A representative field was chosen in which the number of crystals with different morphology was counted. These numbers are reported as a percent of the total number. The experimental conditions and results are listed in Table 5.9 and plotted in Figure 5.15.

Figure 5.15 shows that at low solubility the octahedral morphology is dominating. Beginning at about the log solubility of −5.5, tetrahedral crystals appear that are assigned to a single-twinned morphology.

Beyond −4.5 log solubility, no more octahedra were observed. At about −5.0 log solubility, flat crystals of hexagonal and triangular shape were observed that are

TABLE 5.9
Silver Bromide Morphology as a Function of Solubility

No.	pAg	Solubility (mole/l)	r*/r	Octahedra (%)	Single twin (%)	Double twin (%)	Multiple twin (%)
1	8.00	8.80E-07	0.68	100			
2	8.37	1.30E-06	0.78	100			
3	8.55	1.75E-06	0.79	100			
4	8.78	2.60E-06	0.80	99	1		
5	8.88	3.40E-06	0.84	97	3		
6	9.03	5.00E-06	0.87	90	10		
7	9.19	7.80E-06	0.90	70	27	2	
8	9.32	1.02E-05		56	42	3	
9	9.47	1.90E-05		33	46	20	1
10	9.62	3.10E-05			10	76	17
11	9.69	4.20E-05				76	21
12	9.76	5.50E-05				73	27
13	9.91	1.00E-04				64	36

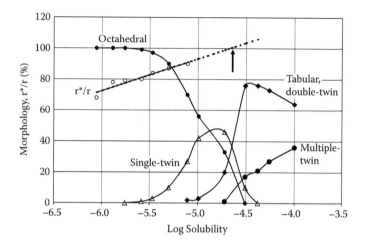

FIGURE 5.15   Silver bromide crystal morphology as a function of solubility.

assigned to double-twinned morphology. This increase in the proportion of double-twinned crystals is accompanied by the decrease of the fraction of single-twinned crystals. After −4.3 log solubility, no more single twins are observed.

At about −4.7 log solubility, large multifaceted tridimensional crystals appear whose morphology is considered representing multiple-twinning. At higher solubility, the multi-twins replace the presence of single- and dual-twin crystals.

For reference, the ratio of $100 \times (r^*/r)$, noted as % r^*/r, is plotted versus log solubility. Extrapolation of the linear correlation shows a crossover at the limiting 100% r^*/r value at about −4.7 log solubility. This threshold is in good agreement with the log solubility of about −4.5 where octahedra are no longer observed. The small difference between −4.5 and −4.7 log solubility is within the limit of error of experiment as well as the uncertainty of the constants used for the calculation of r^*/r.

This is thought to be the first time that a nucleation theory quantitatively predicted the breakdown of a crystal morphology system and the resulting replacement by other morphologies. It is another confirmation, together with the predictions for the kinetically controlled nucleation system, of the power of the BNG nucleation and growth model.

TEMPERATURE

Varying the reaction temperature during the nucleation of the crystals is a simple and effective procedure to vary the crystal size. Because the solubility is a function of temperature, the temperature effect has been generally confounded with solubility changes. However, as Equation (5.35) shows, there is a temperature effect in addition to the solubility and its temperature effect. Equation (5.35) appears to indicate that the number of crystals increases with temperature, which contradicts the generally observed decrease of crystal numbers with increasing temperature. As shown in the earlier section (Experimental: Separation of Variables R, C_s, and T), the temperature dependence of the diffusion constant D must be included.[5,6,8]

$$Z = \left[\frac{R_g}{2k_v \gamma D V_m} \right] \frac{RT}{C_s (r/r^* - 1.0)} \qquad (5.35)$$

$$\frac{d\ln Z}{d\ln T} = -\frac{T}{r/r^* - 1.0} * \frac{d(r/r^*)}{d\ln T} \qquad (5.36)$$

To determine the temperature apart from its effect on solubility, the solubility must be held constant while the temperature is varied. Under these conditions, the balanced nucleation and growth model predicts the temperature dependence of nucleation as shown in Equation (5.18).

As for solubility, the two branches of the solubility curve must be considered for the temperature effect on nucleation. For silver bromide, this is evident by the change of morphology from cubic at low pAg to octahedral at pAg above the minimum of solubility. For silver chloride, cubic crystals are obtained throughout the full pAg range.

SILVER CHLORIDE

The precipitation conditions and experimental results for silver chloride are shown in Table 5.10. The pAg was adjusted at each temperature so that equal solubility was obtained.

The log number of crystals was plotted versus log temperature in Figure 5.16 as suggested by Equation (5.18). As predicted, the log number decreases with increasing log temperature. In addition, the slope of the correlation is higher for experiments at pAg lower than the minimum solubility. For pAg greater than the minimum solubility, the slope is significantly lower.

These differences are significant for the crystallization of silver chloride. If a strong temperature effect is desired, it is preferential to choose the lower pAg range. If greater robustness of crystal number versus temperature effects is desired, the higher pAg range would be preferred.

In addition to the pure temperature effect, the temperature dependence of solubility must be considered. The plot of silver chloride crystal size versus log temperature in Figure 5.17 reflects the results of Figure 5.16. For the control of size at high pAg, the variation of solubility appears more effective than changing temperature. On the other hand, small temperature variations may not introduce greater size variability than predicted, owing to the temperature effect on solubility.

The ratio r^*/r is also greater in the lower pAg range than for the higher pAg range. For the higher pAg range, the ratio decreases significantly with log temperature, while for the lower pAg range, the ratio increases somewhat with log temperature. As was shown in the previous section for silver bromide, higher r^*/r ratios lead to increased instability of the precipitations.

TABLE 5.10
Silver Chloride Nucleation Dependence on Temperature

No.	Temperature (°C)	pAg	Size (µm)	Number (Z)	r*/r
pAg<=5.7					
1	35	4.95	0.115	3.39E+15	0.88
2	40	4.96	0.129	2.44E+15	0.86
3	45	4.98	0.151	1.52E+15	0.82
4	50	5.00	0.173	1.00E+15	0.76
5	55	5.02	0.184	8.32E+14	0.74
6	60	5.10	0.217	5.06E+14	0.66
7	70	5.20	0.235	4.00E+14	0.65
pAg >5.7					
1	35	8.62	0.213	5.29E+14	0.55
2	40	8.23	0.215	5.21E+14	0.57
2	45	7.97	0.211	5.55E+14	0.62
3	50	7.70	0.220	4.99E+14	0.61
3	55	7.40	0.231	4.23E+14	0.60
4	60	7.11	0.235	4.00E+14	0.60
4	70	6.60	0.235	4.00E+14	0.65
5	80	5.68	0.248	3.39E+14	0.64

Temperature, 60°C; addition rate, 0.010 mole/min; initial reactor volume, 0.75 l; reactant flow rate, 10 ml/min; reaction time, 20 min; morphology: cubic

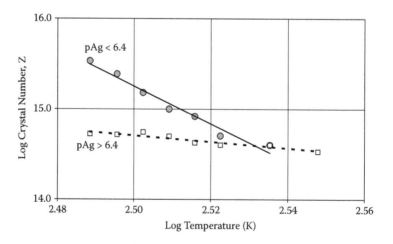

FIGURE 5.16   Silver chloride log crystal number as a function of log temperature.

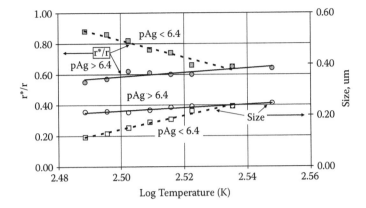

FIGURE 5.17   Silver chloride size and r*/r as a function of log temperature.

SILVER BROMIDE

The experimental details and results for silver bromide are listed in Table 5.11. The results are ordered by temperature and by morphology of the crystals.

The plot of log crystal number as a function of log temperature indicates that the number of crystals decreases with temperature at approximately the same slope for cubic (low pAg range) and octahedral crystals (high pAg range, Figure 5.18). This is in agreement with the prediction of Equation (5.18).

Figure 5.19 shows that r*/r decreases with increasing temperature for both cubic and octahedral precipitation ranges. The slopes of the plots also show a parallel decrease.

For the silver bromide crystal sizes, the temperature dependence is somewhat stronger for octahedral than for cubic crystals. Altogether, these differences are relatively small.

TABLE 5.11
Silver Bromide Nucleation Dependence on Temperature

No.	Temperature (°C)	pAg	Size (μm)	Number (Z)	r*/r	Morphology
1	40	10.2	0.090	5.39E+16	0.86	octahedra
2	50	9.6	0.107	3.23E+16	0.82	octahedra
3	60	9.1	0.126	1.97E+16	0.78	octahedra
4	70	8.4	0.136	1.58E+16	0.77	octahedra
1	40	6.5	0.097	2.04E+16	0.71	cubic
2	50	6.5	0.111	1.35E+16	0.67	cubic
3	60	6.6	0.126	9.24E+15	0.62	cubic
4	70	6.7	0.138	7.08E+15	0.60	cubic

Temperature, 70°C; addition rate, 0.032 mole/min; initial reactor volume, 1.5 l; reaction time, 20 min

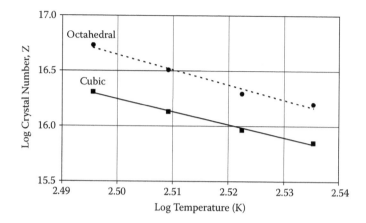

FIGURE 5.18 Silver bromide log crystal number as a function of log temperature.

In conclusion, if the temperature effect on nucleation is determined for constant solubility, a decrease of crystal number with increasing temperature was predicted and was experimentally observed for silver chloride and bromide precipitations. The temperature dependence may vary significantly for different precipitation conditions as shown for silver chloride. The negative slope of r*/r versus log temperature is in agreement with a positive slope for the ratio of r/r* versus log temperature, and the predicted and observed negative slope for the log number/log temperature correlation.

For fundamental studies of the temperature effect on the results of precipitations, it cannot be assumed that the temperature only derives from an increase of solubility. The direct temperature effect must also be considered. For precipitations that are mainly directed for practical control of crystal size, the confounding of the direct

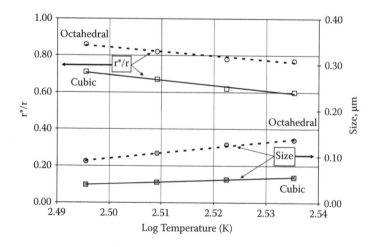

FIGURE 5.19 Silver bromide size and r*/r as a function of log temperature.

effect and the solubility part of temperature may be tolerated. It still is anticipated to lead to a linear predicted log crystal number versus log temperature plot for precision control of crystallization.

In conclusion, the BNG model separates the direct effect of temperature and the solubility for the effect on crystal nucleation. The present chapter confirms predicted direct temperature effect.

REFERENCES

1. Chateau, H., J. Pouradier, and C. R. Berry. 1971. In *The theory of the photographic process*. Ed. T. H. James. 3rd ed. New York: Macmillan.
2. Pouradier, J., A. Pailliotet, and C. R. Berry. 1977. In *The theory of the photographic process*. Ed. T. H. James. 4th ed. New York: Macmillan.
3. Lide, D. R., ed. 2004. *Handbook of chemistry and physics*. Boca Raton, FL: CRC Press.
4. Strong, R. W., and J. S. Wey. 1979. *Photogr Sci Eng* 23:344.
5. Leubner, I. H., R. Jagannathan, and J. S. Wey. 1980. Formation of silver bromide crystals in double-jet precipitations. *Photogr Sci Eng* 24:26–72.
6. Leubner, I. H. 1985. Formation of silver halide crystals in doublejet precipitations: AgCl. *J Imaging Sci* 29:219–25.
7. Leubner, I. H. 1987. Crystal formation (nucleation) under kinetically controlled and diffusion-controlled growth conditions. *J Phys Chem* 91:6069–73.
8. Wikipedia. http://en.wikipedia.org/wiki/Viscosity
9. Wikipedia. 2008. http://en.wikipedia.org/wiki/Surface_tension
10. www.nikhef.nl/~h73/knlc/praktikum/phywe/LEP/Experim/1_4_05.pdf
11. Adam, Neil Kensington. 1941. *The physics and chemistry of surfaces*. 3rd ed. Oxford University Press.
12. Antoniadis, M. G., and J. S. Wey. 1993. *J Imaging Sci Technol* 37:272.
13. Leubner, I. H. 1993. Number and size of AgBr crystals as a function of addition rate. A theoretical and experimental review. *J Imaging Sci Technol* 37:510.
14. King, T. W., W. M. Shor, and D. A. Pitt. 1981. *Photogr Sci Eng* 25:70.

6 Supersizing with Ripeners

CRYSTAL FORMATION (NUCLEATION) IN THE PRESENCE OF OSTWALD RIPENING AGENTS: THEORY

SUMMARY

In the presence of Ostwald ripening agents ("ripeners"), crystal nucleation yields fewer but larger crystals than in their absence. The balanced nucleation and growth model of crystal formation (nucleation) was used to predict the effect of Ostwald ripening agents on crystal nucleation. The model predicts that below a limiting concentration, nucleation is independent of ripener concentration. Above this concentration, in the active ripener range, the number of crystals, Z, decreases with increasing ripener concentration, R_0, while the size increases. In this concentration range, a linear log Z/log R_0 decrease is predicted and observed. For otherwise equal precipitations, this translates to a linear log size/log R_0 increase. The predictions were experimentally verified by double-jet precipitations of cubic and octahedral silver bromide crystals in the presence of an Ostwald ripening agent, 1,10-dithia-4,7,13,16-tetraoxacyclooctadecane, ammonia, and a 1,8-dihydroxy-3,6-dithiaoctane. These correlations allow precise enlargements (super-sizing) of crystal size.

INTRODUCTION

In silver halide precipitations, fewer crystals are formed when certain compounds are present during the crystal nucleation phase compared to when they are not present. This is apparent in significantly increased crystal sizes relative to control precipitations without such addenda (Figure 6.1). It was also observed that these addenda act as silver halide solvents and Ostwald ripeners (Gibbs Thompson effect) due to soluble silver complex formation.[1]

Ostwald ripening is the growth of large crystals by dissolution and reprecipitation of small crystals. This effect is enhanced by certain compounds, Ostwald ripeners, which in the case of silver halides increase the silver halide solubility.

Ostwald ripening may also cause a decrease in crystal number after nucleation during the growth period and after the end of precipitation. In the present experiments, the crystal number stayed constant after the first 3 minutes into the precipitation when the samples were taken. This indicates that the effect of the ripener was restricted to the nucleation step. Until this work was published, no quantitative correlation between ripener properties and concentration, the resultant reduction in crystal number, or designed experiments had been reported.[2]

It was believed that the BNG theory should also be able to predict the effect of ripeners on the nucleation. In this section, a quantitative model for the ripener effect on nucleation will be derived. Experimental support will be shown in the following sections.[3-6]

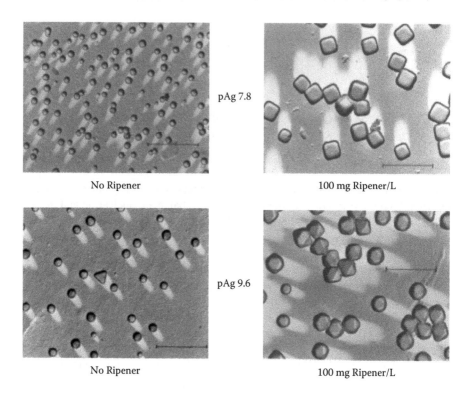

	pAg 7.8	

No Ripener   100 mg Ripener/L

pAg 9.6

No Ripener   100 mg Ripener/L

FIGURE 6.1   Effect of ripener on the crystal size of AgBr at pAg 7.8 and 9.6.

THEORY

The theory of homogeneous nucleation under diffusion-controlled growth conditions leads to Equation (6.1), which was experimentally verified for silver bromide
and silver chloride precipitations.[2–5]

$$Z = \left[\frac{R_g}{2k_v \gamma D V_m} \right] \frac{RT}{C_s (r/r^* - 1.0)} \qquad (6.1)$$

Here, Z is the number of stable crystals formed. The reaction variables are R, addition rate (moles/sec); C_s, solubility (mole/cm^3); and T, absolute temperature (K).
Independent variables are k_v, volume conversion factor, which translates the crystal
size into crystal volume; γ, surface energy (erg/cm^2); R_g, gas constant; D, diffusion
coefficient (cm^2/sec); V_m, molar volume (cm^3/mole); r, average crystal size; and r*,
the critical crystal size. The critical crystal size, r*, is the crystal size at which a
crystal has equal probability of dissolving or growing to a larger size in the system
under investigation. The factor (r/r*–1.0) is the only unknown in Equation (6.1) and
is determined from the experiment.

   We start with the hypothesis that ripeners affect the nucleation process by increasing the silver halide solubility, C_s, in the system. For the condition where all other

variables are held constant, the dependence of crystal number, Z, on solubility is given by Equation (6.2).

$$\frac{d\ln Z}{d\ln C_s} = -\left[1.0 + \frac{C_s}{(r/r^* - 1.0)} * \frac{d(r/r^*)}{dC_s}\right] \quad (6.2)$$

It was experimentally determined that for the silver bromide and silver chloride systems, the derivative $d\ln Z/d\ln C_s$ was a negative constant. We can thus replace Equation (6.2) with Equation (6.3), where k_s is a constant relating the change of crystal number to a change in silver halide solubility.

$$\frac{d\ln Z}{d\ln C_s} = -k_s \quad (6.3)$$

This equation can be solved for Z, and we obtain Equation (6.4) where a is integration constant.

$$Z = aC_s^{-k_s} \quad (6.4)$$

The constant, k_s, was found to depend on the silver halide and, for silver bromide, also on the crystal morphology.

The silver ion solubility C_s in the presence of ripeners is the sum of the concentrations of the silver halide complexes, C_0, and of the silver ripener complexes, C_R (Equation [6.5]). If mixed silver halide/ripener complexes are formed, their concentration, $C_{0,R}$, must be included also. For the present, we will consider the concentration of such mixed complexes to be low versus the other concentrations. Therefore

$$C_s = C_0 + C_R \quad (6.5)$$

The concentration of the silver halide complexes is given by Equation (6.6). The complexing constants necessary to calculate the silver halide solubility as a function of halide concentration and temperature have been published and are listed in the section on solubility in Chapter 5.[1] The brackets in the equations represent the activities of the ions in solution, i.e., the concentrations multiplied by the activity coefficients. In dilute solutions, the activity coefficients are essentially equal to 1.

$$C_0 = [Ag^+] + \sum_{n=1}^{4} [AgX_n]^{1-n} \quad (6.6)$$

Silver ions and ripeners may form a variety of complexes that can contribute to the silver ion solubility (Equation [6.7]).

$$Ag^+ + nR \rightleftarrows [AgR_n]^+, \quad n = 1-4 \quad (6.7)$$

Complexes with several Ag-ions (multinuclear, Ag_mR_n) have been indicated but have not been determined for common silver ion ligands for the present precipitation conditions.[1] Thus, for the remainder of this treatment we will assume mononuclear complexes (m = 1). The fundamental aspects of the theoretical development are not significantly affected by this simplification, and they can be added to the derivation if needed.

From this follows that the silver ripener complex concentration is the sum of the concentrations of the individual complexes (Equation [6.8]).

$$C_R = \sum_{n=1}^{4} AgR_n^+$$

(6.8)

The concentration of the individual silver ripener complexes, AgR_n, can be calculated using Equation (6.9) where K_R is the equilibrium constant.

$$\sum_{n=1}^{4} AgR_n^+ = \sum_{n=1}^{4} K_{R,n}[Ag^+]*[R^n]$$

(6.9)

Values of K_R and of enthalpy changes for silver complex formation in water have been reported for many silver complexing compounds.[1] Thus, the silver complex concentration, C_R (= sum of AgR_n), can be calculated as a function of pAg and temperature and can be inserted into Equations (6.4) and (6.5). However, for the experimental correlation of crystal number with ripener concentration, this is not necessary.

We will now determine in more detail how the decrease in crystal number depends on the ripener concentration, since this is the experimentally simplest method to test the effect of the ripener.

1. Low ripener concentration: $C_0 > C_R$. At low ripener concentrations, where C_R is significantly smaller than C_0, the intrinsic silver halide solubility, the crystal number is independent of the ripener concentration. Only when C_R becomes comparable or significantly larger than C_0 will the ripener significantly affect the crystal number.
2. High ripener concentration: $C_R > C_0$. We will now derive the correlation between number of stable crystals and total ripener concentration, R_0, when the silver-ripener complex concentration C_R is significantly greater than the intrinsic silver halide solubility, C_0. Equation (6.9) contains as one of the unknowns the concentration of noncomplexed ripener, R. We can express this unknown by the mass balance, Equation (6.10).

$$[R] = R_0 - \sum_{n=1}^{4} n[AgR_n^+]$$

(6.10)

The sum of the silver-ripener complex concentrations, C_R, can be calculated when $K_{R,n}$, Ag^+, and R_0 are known (e.g., see Table 6.1).

To determine the concentration dependence of crystal number, Z, on ripener concentration, R_0, the theoretical treatment can be simplified when the silver ion concentration, Ag^+, is held constant, which simultaneously fixes the coordination number, n. For a given precipitation condition, i.e., constant temperature and silver

TABLE 6.1

AgBr Crystal Number and Size as a Function of DTOD Ripener Concentration

	Cubic			Octahedral	
Mg/L	L (μm)	N × 10^{15}	Mg/L	L (μm)	N × 10^{15}
0	0.070	54.200	0	0.137	15.400
10	0.093	23.100	10	0.139	14.700
30	0.130	8.470	30	0.137	15.400
100	0.259	1.070	100	0.236	3.010
200	0.372	0.361	300	0.392	0.657
500	0.432	0.231	500	0.460	0.406
1000	0.637	0.072			

ion concentration, we can introduce the approximation that the ratio of complexed to free ripener is a constant, f (Equation [6.11]).

$$f = n[AgR_n^+]/R_o \qquad (6.11)$$

We can now back-substitute Equation (6.11) into Equations (6.10), (6.9), (6.8), and (6.4) with the condition that $C_R \gg C_0$, and then take the logarithm of Z to arrive at Equation (6.12).

$$\text{Log } Z = -(nk_s)\log R_0 + \log (aK) \qquad (6.12)$$

where

$$K = K_{R,n}[Ag^+](1.0-f)^n \qquad (6.13)$$

The correlation between number of crystals and ripener concentrations becomes now very simple when Ag^+ and n are fixed. At very low ripener concentrations, the number of crystals is constant and does not depend on the ripener concentration. At high ripener concentrations, log Z decreases linearly with increasing log R_0.

In the following sections, experimental results will be presented for the precipitation in the presence of ripener to support the predictions of the model.

CRYSTAL FORMATION (NUCLEATION) OF SILVER BROMIDE IN THE PRESENCE OF A DI-THIA CROWN ETHER AS AN OSTWALD RIPENING AGENT

INTRODUCTION

In the presence of Ostwald ripening agents, crystal nucleation yields fewer but larger crystals than in their absence. The balanced nucleation and growth model of crystal formation (nucleation) was used to predict and quantify the effect of Ostwald ripening agents ("ripeners") on crystal nucleation.[2]

It is known that crown ethers may act as metal complexing agents. Silver ions are efficiently complexed by crown ethers where the ether junctions are replaced by thia-bridges. Such a complex-forming di-thia crown ether was studied for its ripener effect during nucleation for the precipitation of silver bromide.

The model predicts that nucleation is invariant below a limiting (critical) ripener concentration, C_L, and is equal to that of a precipitation without ripener present (Equation [6.14]).

$$Z_R = Z_0 \text{ for } C_R < C_L \tag{6.14}$$

$$Z_R < Z_0 \quad \text{for} \quad C_R > C_L \tag{6.15}$$

$$\text{Log } Z = -(nk_s)\log R_0 + \log (aK) \tag{6.16}$$

The model and equation predict that above this concentration, in the active ripener range, the number of crystals, Z, decreases with increasing ripener concentration, R_0 (Equation [6.15]). In this concentration range, a linear decrease of log Z/log R_0 plots is predicted (Equation [6.16]). For otherwise equal precipitations, this translates to a linear increase of the log Size/log R_0 correlation. These correlations allow precise enlargements (super-sizing) of crystal size.

The predictions were experimentally verified by double-jet precipitations of cubic and octahedral silver bromide crystals in the presence of an Ostwald ripening agent, 1,10-dithia-4,7,13,16-tetraoxacyclooctadecane (DTOD, Figure 6.2).

The correlation between number of crystals and ripener concentrations becomes now very simple when Ag^+ and n are fixed. At very low ripener concentrations, the number of crystals is constant and does not depend on the ripener concentration. At high ripener concentrations, log Z decreases linearly with increasing log R_0.

In the following section, experimental results will be presented for the precipitation of AgBr in the presence of ripener to support the predictions of the model.

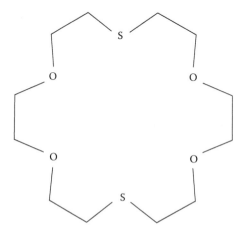

FIGURE 6.2 Molecular structure of ripener 1,10-dithia-4,7,13,16-tetraoxacyclooctadecane (DTOD).

EXPERIMENTAL

The ripener, 1,10-dithia-4,7,13,16-tetraoxacyclooctadecane (DTOD), was chosen as an example to test the ripener model.[7] It forms a 1:1 complex with silver ions.

The $\log(K_R)$ of the silver complex was determined to 9.4 and 7.7 for 25° and 70°C, respectively.[8] From these data a $\log K_R$ value of 8.75 was calculated for the present precipitation temperature of 50°C.[8] Silver bromide was precipitated at pAg 7.8 (cubic morphology, $C_0 = 0.23 \times 10^{-6}$ mole Ag^+/L) and also at pAg 9.6 (octahedral morphology, $C_0 = 1.6 \times 10^{-6}$ mole Ag^+/L).

Sixty grams of deionized dry bone gel were dissolved in 1.5 L distilled water and heated to 50°C. The ripener was added, dissolved in 25 mL methanol. After the pAg stabilized, NaBr was added to adjust to the desired pAg. $AgNO_3$ and NaBr solutions (0.032 mole/min for 20 min) were added in a controlled double-jet precipitation. The initial ripener concentration in the precipitations was varied from 0 to 1,000 mg/L.

The crystal sizes were determined from electron micrographs. The number of crystals was calculated from the crystal size and mass balance. The crystal number and sizes are summarized in Table 6.1.

RESULTS AND DISCUSSION

Crystal Number

The experimental results were plotted in Figure 6.3 according to the predictions of the model, i.e., log (crystal number, Z) versus log (ripener concentration). The crystal numbers of the precipitations without ripener are indicated by lines parallel to the concentration axis.

The experimental results are in excellent agreement with the predictions of the model. Above a limiting (critical) ripener concentration, the number of crystals decreases with increasing ripener concentration, and the log Z/ log R_0 correlation is linear.

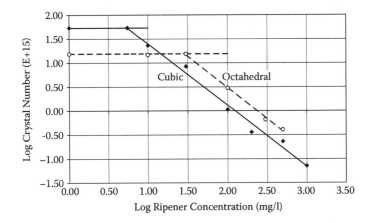

FIGURE 6.3 Log (number) versus log (R_0) for AgBr precipitations at pAg 7.8 (cubic crystals) and pAg 9.6 (octahedral crystals).

The slopes of the curves (cubic: −1.27 for R_0 > 5.5 mg/L; octahedral: −1.31 for R_0 > 30 mg/L) for the octahedral and cubic AgBr precipitation conditions are essentially the same. This is expected for a ripener that does not change coordination number as a function of silver ion concentration.

At low concentrations, a limiting ripener concentration is indicated below which the ripener does not affect the crystal number of the precipitation. This limiting ripener concentration is lower for the cubic (pAg 7.8) than for the octahedral (pAg 9.6) precipitation conditions. This is in agreement with the magnitude of intrinsic silver halide solubilities as indicated in Equation (6.18) (pAg 7.8: $C_0 = 0.23 \times 10^{-6}$ mole Ag/L; pAg 9.6: $C_0 = 1.6 \times 10^{-6}$ mole Ag/L) and the greater silver-ripener-complex concentration at pAg 7.8 (0.9 R_0) versus pAg 9.6 (0.124 R_0). The actual offset in the crystal number/ripener concentration correlations (Figure 6.3) may also be affected by the change from cubic to octahedral morphology of the crystals, as indicated by the Z/C_0 correlations shown in the section on solubility in Chapter 5.

CRYSTAL SIZE

The crystal size is related to the number of crystals and the total mass of crystals formed by Equation (6.17). The constant, k_v, is the volume shape factor that relates crystal shape to crystal volume.

$$r^3 = \frac{Z}{k_v V_m \int Rdt} \tag{6.17}$$

For identical precipitations and equal crystal shape, Equation (6.17) can be rewritten to Equation (6.18).

$$\log r = \frac{1}{3}\log Z \tag{6.18}$$

Under these conditions, a plot of log r versus log ripener concentration should lead to a linear correlation in the active region with a slope of 1/3 of the log Z correlation (Equation [6.19]).

$$\log r = -(nk_s/3)\log R_0 + \log (aK)/3 \tag{6.19}$$

The results of the log r versus log R_0 plot are shown in Figure 6.4. Since for the size the edge length of the crystals was used, the overlap of the data does not reflect different correlations of the crystal numbers.

CONCLUSIONS

The nucleation model for homogeneous nucleation under diffusion-controlled growth conditions was extended to predict the effect of ripening agents on the number of crystals formed.

The model predicts that above a limiting ripener concentration, the number of crystals will decrease. A linear correlation for a plot of log (crystal number) versus

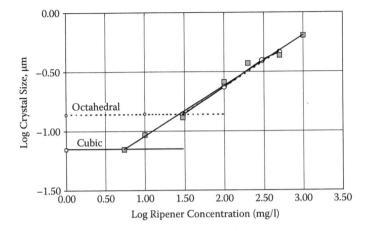

FIGURE 6.4 Log (size) versus log (R_0) for AgBr precipitations at pAg 7.8 (cubic crystals) and pAg 9.6 (octahedral crystals).

log (ripener concentration) was predicted. Below the limiting ripener concentration, the crystal number is determined by the silver halide solubility.

The predictions of the theory were tested by silver bromide precipitations under fixed precipitation conditions where only the ripener concentration was changed. Precipitation conditions were chosen to form cubic and octahedral crystals, respectively. The predicted correlations were obtained for both morphologies.

Plots of the crystal size, r, for otherwise equal precipitations, result in the predicted linear increase of the log (r)/log (R_0) correlation, with, however, one-third the slope of the log (Z)/log (R_0) correlation.

The model and the experiments results indicate that a minimum of three experiments is necessary to define the ripener effect in a given system: (a) one control precipitation without ripener and (b) two precipitations in the active ripener concentration range. Additional variations of ripener concentration will make the correlation more precise and improve the predictability of the model.

In conclusion, the use of ripeners during the nucleation stage of crystallizations may provide a powerful method of precisely increasing the crystal size without the change of other precipitation conditions, and with the retention of the crystal morphology.

CRYSTAL FORMATION (NUCLEATION) OF AgClBr (15:85) IN THE PRESENCE OF AMMONIA AS AN OSTWALD RIPENING AGENT

INTRODUCTION

It is generally known that ammonia at high concentrations will dissolve silver halides and at low concentrations cause accelerated growth of small crystals to larger ones (Ostwald ripening). Thus, ammonia has the properties of a ripening agent. The BNG theory of the ripening effect predicts that during the nucleation phase ammonia should cause a decrease of nuclei and an increase in the crystal size. In addition, the

BNG model predicts a linear correlation of the crystal number and size as a function of log ripener concentration above a limiting concentration. These predictions were tested for precipitations of silver chlorobromide in the presence of ammonia.

EXPERIMENTAL

This prediction was tested by adding ammonia to the precipitation medium before crystallization was started. The concentration was varied from zero (reference) to 300 mmole/l.

The precipitations were performed with a starting reaction volume of 1.5 liter of a 4% gelatin solution. The temperature was adjusted to 40° and 60°C, respectively. When these temperatures had stabilized, the pAg was adjusted with NaCl to 6.89 and 6.44, respectively. An aliquot of aqueous ammonia was added to achieve the desired ammonia concentration in the reactor. The pH and pAg were measured at this point, and the precipitation was started.

The precipitation was begun with a reactant flow rate of 5 ml/min and held constant for 5 minutes. At this time, the flow rate was increased linearly to 45 ml/min. The reaction was ended 45 minutes after the beginning of the precipitation. The ammonia was at once neutralized with diluted sulfuric acid to pH 7.0. Samples were taken and analyzed. Total: 1.54 mole AgClBr (15:85), corresponding to 43.87 cm^3.

RESULTS

Crystal Number and Size

The results of the precipitations are listed in Table 6.2 (40°C) and Table 6.3 (60°C). The log crystal numbers and sizes are plotted versus log ammonia concentration in Figures 6.5 and 6.6 for 40° and 60°C, respectively.

TABLE 6.2
Silver Chlorobromide (15:85) Precipitation with Ammonia as a Ripener at 408C

No.	mMNH$_3$	pH	pAg	Size	Number
1	0.0	5.60	6.89	0.197	5.77E+15
2	4.0	7.90	6.99	0.207	4.97E+15
3	4.0	7.94	7.04	0.187	6.74E+15
4	8.0	8.73	7.20	0.295	1.70E+15
5	8.0	8.71	7.17	0.283	1.94E+15
6	15.0	9.14	7.39	0.325	1.28E+15
7	31.0	9.48	7.50	0.470	4.23E+14
8	31.0	9.48	7.50	0.352	1.01E+15
9	62.0	9.70	7.94	0.573	2.33E+14
10	124.0	9.95	8.23	0.857	6.98E+13
11	300.0	10.33	8.65	1.474	1.37E+13
12	300.0	10.30	8.60	1.378	1.68E+13

TABLE 6.3

Silver Chlorobromide (15:85) Precipitation with Ammonia as a Ripener at 608C

No.#	mMNH$_3$	pH	pAg	Size	Number
1	0	5.60	6.44	0.282	1.96E+15
2	0	5.60	6.44	0.300	1.62E+15
3	0	5.60	6.45	0.282	1.96E+15
4	4	7.50	6.48	0.322	1.31E+15
5	4	7.36	6.48	0.343	1.09E+15
6	8	8.22	6.27	0.479	4.00E+14
7	15	8.56	6.77	0.607	1.96E+14
8	31	8.91	6.97	0.885	6.33E+13
9	31	8.89	6.95	0.705	1.25E+14
10	31	8.89	6.95	0.717	1.19E+14
11	40	8.93	7.08	0.620	1.84E+14
12	40	8.91	7.03	0.715	1.20E+14
13	40	8.92	7.04	0.684	1.37E+14
14	41	9.03	7.08	0.768	9.67E+13
15	62	9.18	7.23	0.832	7.63E+13
16	90	9.20	7.36	1.042	3.88E+13
17	124	9.55	7.51	1.147	2.91E+13
18	300	9.82	7.94	1.683	9.20E+12

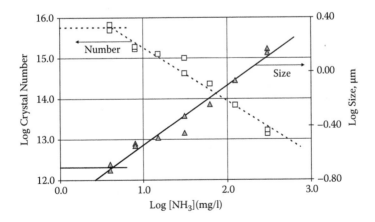

FIGURE 6.5 AgClBr (15:85): 40°C, log crystal number, and size versus ammonia concentration.

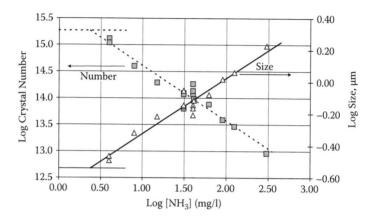

FIGURE 6.6   AgClBr (15:85): 60°C, log crystal number, and size versus ammonia concentration.

The plots show that the predicted correlations apply to the present precipitation conditions. Fortuitously, for the 40°C precipitation, the crystal numbers and sizes at the lowest concentration (4.0 mmole NH_3/l) are at the crossover point with the results for the control precipitation. For the 60°C precipitation, a lower crystal number and a larger crystal size are obtained than for the control precipitation.

In Equations (6.20), (6.21), (6.22), and (6.23), the linear least squares correlations are shown for the crystal numbers and sizes for 40° and 60°C, respectively. The standard deviations and the squares of the Pearson product moment correlation coefficient indicate excellent correlations.

$$40°C \ logZ = -(1.325+/-0.079)* \ log[NH_3]+(16.596+/-0.126) \quad (6.20)$$

$$Log \ Sze = (0.442+/-0.026)*log[NH_3]-(0.985+/-0.042)$$
$$R^2 = 0.9692 \quad (6.21)$$

$$60°C \ Log \ Z = -(1.048+/- \ 0.074)*log[NH_3]+(15.664+/-0.116) \quad (6.22)$$

$$Log \ Sze = (0.349+/-0.025)*log \ [NH_3]-(0.674+/-0.039)$$
$$R^2 = 0.9398 \quad (6.23)$$

The results confirm the linear correlation of the logarithm of crystal number and size with the logarithm of the ammonia concentration in the active ripener concentration range.

Close inspection of the plots and tables, however, shows relatively large variability for repeat experiments. The causes for the variability may be manifold. Pipettes were used to measure and apply the ammonia solutions. As one point, the exact addition and manipulation of the ammonia solution is difficult, because ammonia as a gas may evaporate from the solution or the reaction mixture at different rates. Another cause may be inadvertent Ostwald ripening during the precipitation and during handling after the end of precipitation. Although care was taken to closely monitor the

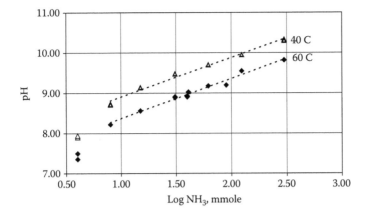

FIGURE 6.7 AgClBr (15:85): pH as a function of ammonia concentration at 40° and 60°C.

manipulations, variability of repeat experiments apparently could not be avoided in the experimental setup. The data thus suggest that the manipulation of gaseous reactants in precipitations needs careful attention and control.

Ammonia Effects in the Reactor on pH

The pAg and pH of the reaction mixture directly after ammonia addition were monitored and recorded. The results are included in Tables 6.2 and 6.3. Correlations between pH and pAg and log ammonia concentration are shown in Figure 6.7 and Figure 6.8, respectively. The reference values before addition of ammonia are pH 6.0 and pAg 6.44.

The plot of pH versus log NH_3 shows that the lowest ammonia concentration (4.0 mmole/l) causes a significant increase of the pH from 5.60 to 7.9 (40°C) and 7.5

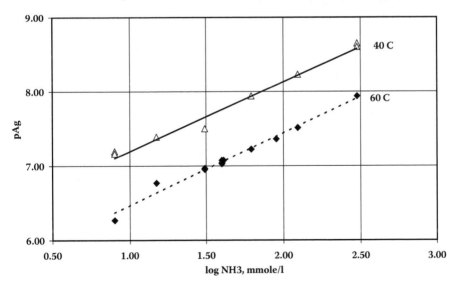

FIGURE 6.8 AgClBr (15:85): pAg as a function of ammonia concentration at 40 and 60°C.

(60°C). The pAg increased less dramatically from pAg 6.89 to 7.02 for 40°C, and from pAg 6.44 to 6.48 at 60°C.

A linear correlation between pH and log NH$_3$ was only observed at and above 8.0 mmole ammonia/l. The correlations for 40° and 60°C are shown in Equations (6.24) (40°C) and (6.25) (60°C).

$$40°C \; pH = (0.981+/-0.045)*logNH_3 + (7.921+/-0.078)$$

$$R^2 = 0.9855 \tag{6.24}$$

$$60°C \; pH = (1.001+/-0.045)*logNH_3 + (7.362+/-0.075)$$

$$R^2 = \mathbf{0.9785} \tag{6.25}$$

The calculated slopes of about 1.0 confirm the parallel correlation in Figure 6.7. The nonlinearity at low concentrations is a warning that such correlations must be experimentally confirmed before using the pH as a reaction variable. In the present case, the initial nonlinearity is a function of the buffering effect of the gelatin in the reaction mixture.

Ammonia Effects in the Reactor on pAg

The tables show that the pAg of the reaction mixture is a function of the ammonia concentration. The correlation between pAg and log ammonia concentration is illustrated in Figure 6.8 for 40° and 60°C, respectively.

The figure shows that the pAg increases linearly with the log ammonia concentration at and above 8.0 mmole/l. The 4.0 mmole data are omitted, since their pAg values are well below the correlation.

Correlations (Equations [6.26] and [6.27]) confirm the results in Figure 6.8 that the pAg is proportional to the logarithm of the ammonia concentration. Further, the correlations for 40° and 60°C are parallel. These correlations indicate that above a limiting value, the pH or pAg may be used to control the ripening effect instead of the ammonia concentration.

$$40°C \; pAg = (0.981+/-0.045)*log \; NH_3 + (7.921+/-0.078)$$

$$R^2 = 0.9855 \tag{6.26}$$

$$60°C \; pAg = (0.982+/-0.040)*logNH_3 + (5.483+/-0.067)$$

$$R^2 = 0.9823 \tag{6.27}$$

CONCLUSION

The precipitation of silver chlorobromide (15:85 molar ratio) in the presence of ammonia confirms the prediction that ammonia acts as a ripener during nucleation.

The results confirm the theoretical predictions that the logarithm of crystal number decreases and the logarithm of crystal size increases with the logarithm ammonia concentration above a limiting critical concentration.

It was observed that the presence of ammonia in the reaction mixture causes an increase of pH and pAg. Above a critical ammonia concentration, the pAg and pH are linearly correlated with the logarithm of the ammonia concentration. The slope of these correlations is independent of temperature as determined for 40° and 60°C. These results indicate that for practical purposes, the monitoring of the pH and pAg may provide control parameters for the effect of ammonia.

CRYSTAL FORMATION (NUCLEATION) OF AgBrI (97.4:2.6) IN THE PRESENCE OF 1,8-DIHYDROXY-3,6-DITHIAOCTANE AS AN OSTWALD RIPENING AGENT

INTRODUCTION

The present experiments supplement the preceding ripener examples in a number of ways. In addition to the variation of ripener concentration, the effect of temperature and of initial pAg was studied in a statistically designed experiment. A different ripening molecule, 1,8-dihydroxy-3,6-dithiaoctane (DTO, CH_2OH-CH_2CH_2-S-CH_2-CH_2-S-CH_2-CH_2OH), is introduced. In addition to acting as a ripener dissolved in the reaction mixture, it adsorbs to the crystal surface. The adsorption was analytically determined. At high concentrations, the adsorption causes changes from cubic to rounded morphology. The crystal material, silver bromoiodide (97.4: 2.6 molar ratio), extends the range of crystal composition.

EXPERIMENTAL

A reactor was charged with 8.5 liters of a 2.4% aqueous solution of gelatin. After the aim temperature was reached, the pAg was adjusted with a sodium bromide solution. The ripener, 1,8-dihydroxy-3,6-dithiaoctane (DTO), was dissolved in 50 ml of methanol and added to the reaction mixture. Aqueous solutions of 2.5 mole/l silver nitrate and of 2.5 mole/l sodium bromide containing potassium iodide (molar ratio 97.4:2.6) were added at a rate of 85 ml/min for 35 min. Five minutes into the precipitation, the pAg was changed from the initial nucleation pAg to the growth pAg (indicated in Tables 6.4 and 6.5) to provide growth conditions for cubic morphology. Nevertheless, the observed morphology varied from cubic (c) to rounded cubes to round, almost spherical. A total of 7.44 mole (215.7 cm^3) AgBrI (97.4:2.6) was obtained. The reaction mixture was washed using ultrafiltration, and the samples for the size measurements were stabilized with a growth restrainer. In addition to size, the amount of retained adsorbed ripener was determined.

For the statistical factorial design, the temperature was varied as 50°, 60°, and 70°C; the range for the ripener concentrations was 170, 340, and 510 mg/l; the pAg

TABLE 6.4
Ripener Experiments: Statistical Design and Results

No.	Temp. (8C)	pAg Nucleation	pAg Growth	Ripener (mg/l)	Size (μm)	Number (Z)	Ripener (% retained)	Morphology
1	50	8.06	7.91	170.0	0.266	1.15E+16	49.7	c
2	50	8.06	7.91	170.0	0.272	1.07E+16	47.3	c
3	50	8.06	7.91	510.0	0.469	2.09E+15	20.3	r
4	50	8.06	7.91	510.0	0.469	2.09E+15	15.0	r
5	50	8.69	7.91	170.0	0.186	3.35E+16	52.7	c
6	50	8.69	7.91	510.0	0.385	3.79E+15	22.1	r
7	50	8.69	7.91	510.0	0.361	4.57E+15	20.8	r
8	60	8.11	7.65	170.0	0.221	2.01E+16	49.0	c
9	60	8.11	7.65	170.0	0.224	1.91E+16	n/a	c
10	60	8.11	7.65	170.0	0.235	1.67E+16	40.5	c
11	60	8.11	7.65	170.0	0.239	1.58E+16	42.9	c
12	60	8.11	7.65	170.0	0.230	1.76E+16	34.2	c
13	60	8.11	7.65	340.0	0.323	6.43E+15	27.7	cr
14	60	8.11	7.65	340.0	0.342	5.39E+15	27.5	cr
15	60	8.11	7.65	510.0	0.399	3.40E+15	19.3	r
16	60	8.11	7.65	510.0	0.399	3.40E+15	20.5	r
17	70	7.56	7.41	170.0	0.306	7.55E+15	38.0	c
18	70	7.56	7.41	510.0	0.478	1.97E+15	15.6	r
19	70	8.15	7.41	170.0	0.251	1.37E+16	46.4	c
20	70	8.15	7.41	510.0	0.402	3.31E+15	18.8	r

values were adjusted so that the potential difference between a silver electrode and a saturated potassium chloride/silver electrode varied as 40, 60, and 80 mV.

For the ripener experiments, the precipitation details and results are listed in Table 6.4. Results for control experiments, where no ripener was used, are compiled in Table 6.5.

TABLE 6.5
Control Experiments without Ripener: Design and Results

No.	Temperature (8C)	pAg Nucleation	pAg Growth	Size (μm)	Number (Z)
1	50	8.06	7.91	0.109	1.67E+17
2	50	8.69	7.91	0.095	2.53E+17
3	60	8.11	7.65	0.107	1.75E+17
4	70	7.56	7.41	0.140	7.86E+16
5	70	8.15	7.41	0.144	7.29E+16
6	70	8.15	7.41	0.161	5.14E+16

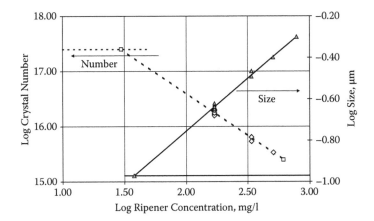

FIGURE 6.9 AgBrI (97.4:2.6): 60°C log crystal number and size versus log ripener concentration.

RESULTS

60°C Results: Crystal Size/Number with Ripener Correlation

To test for the correlation between experimental data and model, the log crystal number and log size results for the factorial center experiments (60°C, pAg 8.11) were plotted versus the log ripener concentration (Figure 6.9).

The figure shows that the plot is in agreement with the predicted linear correlation of log crystal number and size with log ripener concentration. Also shown are the lines parallel to the ripener axes, which represent the size and number of reference experiments without ripener. The quantitative correlations for number and size are shown in Equations (6.28) and (6.29).

$$\log Z = -(1.540 \pm 0.084)\log Rip + (19.682 \pm 0.199)$$

$$R^2 = 0.9825 \tag{6.28}$$

$$\log Sze = (0.5132 \pm 0.0280)\log Rip - (1.783 \pm 0.066)$$

$$R^2 = 0.9825 \tag{6.29}$$

These equations confirm excellent correlations and give confidence to test an overall correlation for all data.

All Data: Size/Number with Ripener Correlation

The statistically designed experiment was anticipated to reveal the main driving control parameters that determine crystal size and number. After the linear correlation for the center experiments (60°C) was established, all experimental number results were plotted in Figure 6.10. The reference lines (no ripener) were omitted in this plot, but the data are given in Table 6.5.

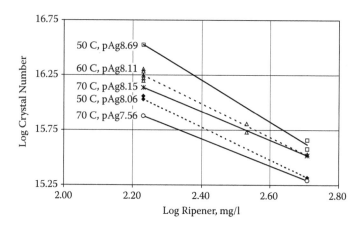

FIGURE 6.10 AgBrI (97.4:2.6): All data, log crystal number and size versus log ripener concentration.

Overall linear correlations are assumed for those data sets where the center ripener concentration (340 mg/l) was not determined experimentally.

A full set of multilinear calculations revealed that ripener concentration and nucleation pAg (N_pAg) were the main driving forces within the experimental range. For temperature and growth pAg, the standard error exceeded the value of the predicted slopes, and these effects were neglected. The results of the calculations using only the ripener concentration and nucleation pAg are shown in Equations (6.30) and (6.31).

$$logZ = -(1.559 \pm 0.091)*logRip + (0.416 \pm 0.073)*N\_pAg + (16.299 \pm 0.609)$$

$$(6.30)$$

$$logSze = (0.520 \pm 0.030)*logRip - (0.139 \pm 0.024)* N\_pAg - (0.655 \pm 0.203)$$
$$(6.31)$$
$$R^2 = 0.9475$$

To visualize the correlation between calculated and experimental data, the experimental and calculated (Equation [6.31]) sizes are plotted in Figure 6.11.

The observed linear correlation is quantified in Equation (6.32). These results indicate that the ripener effect on the resulting crystal size (and number) can be quantified using the proposed ripener model. For the present experiments, a relatively wide range of experiments were done to confirm the correlations and provide a practical predictive tool. This gives confidence that reliable predictions may be made from a smaller exploratory experimental design.

$$Sze\text{-}exp = (1.305 \pm 0.020)*Sze\text{-}calc - (0.430 \pm 0.0113)$$

$$(6.32)$$
$$R^2 = 0.9986$$

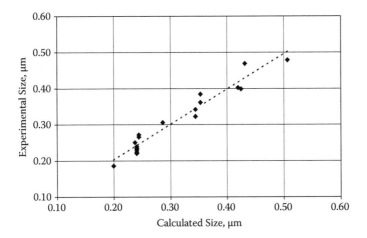

FIGURE 6.11   AgBrI (97.4:2.6): All experiments, experimental versus calculated crystal size.

Retained Ripener

It was determined that ripener was retained at the crystals after precipitation and after ultrafiltration (Table 6.4). The adsorbed ripener may interfere with surface reactions that may be necessary for practical applications, such as catalytic activity, stabilization of the crystals, or to achieve high sensitivity for photographic films.[9]

For statistical evaluation of the results, additional variables were calculated. Thus, it appears reasonable to assume that the total of adsorbed ripener (R_ads, mg) may be a function of (a) the nucleation pAg (N_pAg), since the ripener interacts with silver ions or (b) the total of ripener added (R_add, mg), and the total surface area (AreaTot, m^2; Table 6.6).

The results of the statistical evaluation of all experiments with ripener are shown in Equation (6.33). The standard deviations of the coefficients give high confidence in the model. The relatively low correlation coefficient ($R^2 \sim 0.70$) is suspected to be due to the accumulated experimental errors.

$$R\_ads = (184.6 \pm 1.5)*N\_pAg + (0.0688 \pm 0.0008)*R\_add$$

$$-(0.00128 \pm 0.00009)*AreaTot-(905.7 \pm 13.3) \qquad (6.33)$$

$$R^2 = 0.7040$$

In Figure 6.12, the experimentally obtained ripener is plotted versus the calculated values from Equation (6.33). The scatter of data around the correlation line is an indication that the analytical ripener determination and the reaction control variables contribute unknown errors.

TABLE 6.6
Ripener Adsorption: Experimental Control Factors

No.	Temp. (8C)	pAg Nucleation	Size (μm)	Surface Area (m^2)	Ripener Added (mg)	Ripener Retained (mg)	Ripener Calc. (mg)
1	50	8.06	0.266	3.44E+04	1445.0	718.2	640.8
2	50	8.06	0.272	3.52E+04	1445.0	683.5	639.7
3	50	8.69	0.186	2.41E+04	1445.0	761.5	770.5
4	60	8.11	0.221	2.86E+04	1445.0	708.1	657.5
5	60	8.11	0.224	2.90E+04	1445.0		656.9
6	60	8.11	0.235	3.04E+04	1445.0	585.2	655.2
7	60	8.11	0.239	3.10E+04	1445.0	619.9	654.4
8	60	8.11	0.230	2.98E+04	1445.0	494.2	655.9
9	70	7.56	0.306	3.96E+04	1445.0	549.1	541.7
10	70	8.15	0.251	3.24E+04	1445.0	670.5	659.9
11	60	8.11	0.323	4.17E+04	2890.0	800.5	740.3
12	60	8.11	0.342	4.43E+04	2890.0	794.8	737.1
13	50	8.06	0.469	6.07E+04	4335.0	880.0	806.6
14	50	8.06	0.469	6.07E+04	4335.0	650.3	806.6
15	50	8.69	0.385	4.98E+04	4335.0	958.0	937.1
16	50	8.69	0.361	4.68E+04	4335.0	901.7	940.9
17	60	8.11	0.399	5.16E+04	4335.0	836.7	827.4
18	60	8.11	0.399	5.16E+04	4335.0	888.7	827.4
19	70	7.56	0.478	6.19E+04	4335.0	676.3	712.5
20	70	8.15	0.402	5.21E+04	4335.0	815.0	834.2

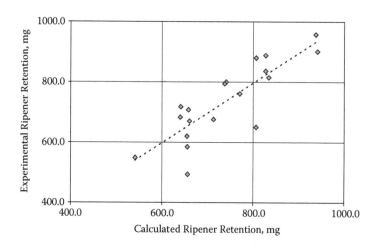

FIGURE 6.12   AgBrI (97.4:2.6): Ripener retention, plot of experimental versus calculated.

SUMMARY

The crystal size and number of controlled AgBrI (97.4:2.6) precipitations was determined in a statistically designed experiment. The fundamental concepts and theoretical predictions of the BNG model were used as a base for the experimental design. Control parameters were temperature, pAg during the nucleation phase, and ripener concentration. To obtain crystals of comparable morphology, the pAg was adjusted to a common value after nucleation. Reactant addition rate and flow rates were not varied.

The statistical evaluation of the crystal size and number as a function of reaction control variables showed that the nucleation pAg and ripener concentration were the major controlling factors. Surprisingly, the reaction temperature was not a leading determining factor.

It was determined that the crystals adsorbed ripener, and the amount of adsorbed ripener could be analytically determined. Statistical evaluation showed that the amount of adsorbed ripener is a function of nucleation pAg, total amount of added ripener, and final surface area of the crystal population. The plot of experimental and calculated ripener retention supported the correlation. However, unknown experimental errors contributed significantly to the scatter of data.

These experiments show that the BNG model provided solid guidance for the experimental design and evaluation of the results. Statistical evaluation of the results allowed determining the controlling reaction variables. At the same time, the statistical design and evaluation, together with the predictions of the BNG model, allowed a significant reduction of the number of experiments while giving substantial information.

REFERENCES

1. James, T. H. 1977. In *The theory of the photographic process*. Ed. T. H. James. 4th ed. New York: Macmillan.
2. Leubner, I. H. 1987. Crystal formation (nucleation) in the presence of Ostwald ripening agents. *J Imaging Sci* 31 (4): 145–48.
3. Leubner, I. H., R. Jagannathan, and J. S. Wey. 1980. *Photogr Sci Eng* 24:268.
4. Leubner, I. H. 1982. *Colloids and surfaces in reprographic technology*. ACS Symposium Series 22. p. 81.
5. Leubner, I. H. 1985. *J Imaging Sci* 29:219.
6. Leubner, I. H. 1987. *J Phys Chem* 91:6069.
7. Dann, J. R., P. P. Chiesa, and J. W. Gates. 1961. *J Org Chem* 26:1991; Frensdorf, H. 1971. *J Am Chem Soc* 27:1; Hengel, R. 1983. *Photogr Sci Eng* 27:1; McBride, C. E. 1966. US Patent 3,271,157.
8. Herz, A., W. L. Gardner, and D. D. F. Shiao. Research Laboratories, Eastman Kodak Company, private communication.
9. Leubner, I. H. 2000. One-photon processes of latent image formation in silver halides. *J Imaging Sci* 47:213.

7 Nanosizing with Restrainers

CRYSTAL FORMATION (NUCLEATION) IN THE PRESENCE OF GROWTH RESTRAINERS

When certain compounds ("restrainers") are present during nucleation of precipitations, more and thus smaller crystals are obtained than in their absence. The model for their mechanism in crystallization is generally applicable and gives general guidance for the controlled synthesis of nano particles.[1] This was experimentally confirmed by the effect of 1-(3-acetamidophenyl)-mercaptotetrazole on the nucleation of silver chloride.

INTRODUCTION

It is the purposes of this chapter to extend the BNG model of crystal nucleation to include the effect of crystal growth restrainers on the crystal nucleation process. The mechanism and quantitative model of crystal growth restrainers is based on the nucleation model for kinetically controlled growth conditions. Experimental results will be presented to support the theoretical model.

The term *restrainer* is used in various contexts in photographic science, e.g., crystal growth or ripening restrainer, restrainer of chemical sensitization, or development restrainer.[2] It is common to all these restrainer actions that the compound under consideration adsorbs to the crystal surface and inhibits surface reactions. In the present context, the concept will be used to explain and model the effect of such compounds on crystal nucleation.

It has been observed that in silver halide precipitations, more crystals are obtained when certain compounds are present during nucleation than when they are absent.[2,3] In practice, the increased number of nuclei is apparent in reduced crystal sizes relative to control precipitations without such addenda (Figure 7.1).

As Figure 7.1 shows, significant changes in crystal morphology and size distribution may also be observed.[3]

Protective colloids, like gelatin, may act as restraining agents, and the effect of gelatin was studied in great detail.[4] Since these compounds generally retard Ostwald ripening of silver halides, they have been classified as growth restrainers.

As an example, the restraining effect of adenine on the nucleation of silver halides was studied. The data suggested that adenine had to adsorb to the surface of the silver halides to be effective. Adsorption took place at the initial stages of precipitation when the crystals presumably still had rather spherical shapes, and the adenine adsorbed preferentially to distinct surface regions that are growth centers. It was found that complete surface coverage of the crystals was not necessary to obtain

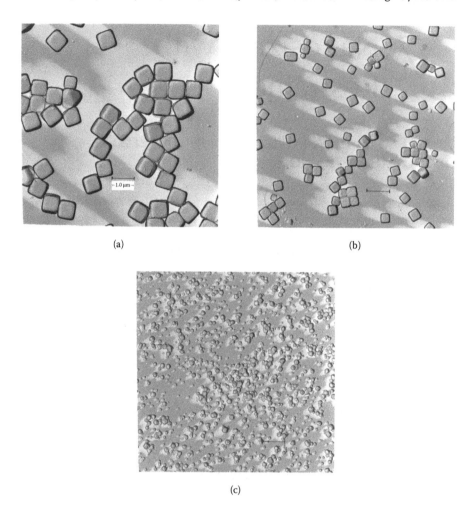

(a)   (b)

(c)

FIGURE 7.1   Electron micrographs of AgCl precipitated without and with restrainer I (see Table 7.1), (a) 0, (b) 10, and (c) 500 mg restrainer/l.

maximum restraining effects. The adsorption depended on the relation of the solubility of the silver adenate to that of the silver halides. Adsorption was most efficient under experimental conditions where the silver adenate was less soluble than the silver halide.

This suggests that adsorption of adenine to the silver halide may proceed through chemisorption. It was postulated that adenine decreases the surface free energy of the crystals and reduces material transport between solid phase and solution proportional to its adsorbed quantity.[3] However, this study did not lead to a quantitative correlation of restrainers on the crystal number and size.

The present model is based on the assumption that the adsorbed restrainer reduces the surface integration of growth material. Previous observations are included in the model to form a quantitative basis for the restrainer effect on crystal nucleation. This leads to a quantitative model that is in agreement with the observed effects of nucle-ation and changes in crystal morphology as a function of restrainer concentration.

MODEL

Surface Integration Model of Nucleation and Effect of Crystal Growth Restrainers

Nucleation and growth of crystals may be limited either by the diffusion of mate-rial to the surface (diffusion-controlled mechanism) or by the surface integration processes (kinetically controlled mechanism), as discussed in Chapters 5 and 4, respectively. For silver bromide and chloride, the crystal growth rates and the crystal nucleation in the presence of gelatin and ripeners are in agreement with diffusion-controlled growth and nucleation models.[5,6] However, in the presence of growth restrainers, interaction with the surface integration processes and thus kinetically controlled growth must be considered.

The presence of adsorbed growth restrainer reduces the crystal growth rate, as modeled by the fundamental BNG model. Figures 7.2 and 7.3 show the effect of adsorbed restrainer, and its effect on maximum growth rate, on nucleation rate and size distribution, respectively.

While the basic model shows the restrainer effect on the nucleation phase, the effect on crystal number and size after nucleation will be quantified using the nucle-ation model under kinetically controlled growth conditions.

In the absence of restrainer, nucleation and growth of AgCl and AgBr is diffusion-controlled. This indicates that kinetic incorporation rate is faster than the material supply by diffusion. Adsorption of restrainer reduces the kinetic incorporation rate, and eventually, transition from diffusion-controlled to kinetically controlled growth is anticipated during the nucleation phase. At further growth after the end of nucle-ation, a reverse transition to diffusion-controlled growth is anticipated for partial surface coverage.

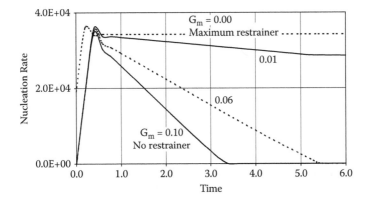

FIGURE 7.2 Nucleation rate as a function of growth rate, G_m.

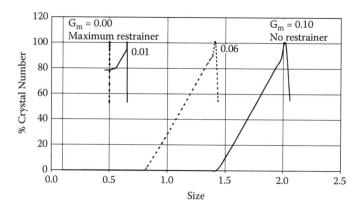

FIGURE 7.3   Size and size distribution as a function of growth rate, G_m.

The derivation of a kinetically controlled model for crystal nucleation leads to Equation (7.1).

$$Z = \frac{RR_g T}{8\pi\gamma C_s K_i r(r/r^* - 1.0)} \tag{7.1}$$

Here, Z is the total number of crystals, r the average crystal size, K_i the surface integration constant, R the addition rate (mol/s), R_g the gas constant, T the absolute temperature, γ the surface energy, C_s the solubility, and r^* the critical crystal size. The critical crystal size, r^*, is defined as the crystal size at which a crystal has equal probability to dissolve or grow in the given reaction environment.

For the case where all reaction variables (addition rate, temperature, and solubility) except the restrainer concentration are held constant, Equation (7.1) can be simplified to Equation (7.2). It is assumed that changes in surface energy and r/r^* due to adsorption are reflected in a change of K_i:

$$Zr = K/K_i \tag{7.2}$$

In the presence of a growth restrainer that adsorbs to the crystal surface, K_i must be substituted by a relative surface integration constant, K_r (Equation [7.3]).

$$Zr = K/K_r \tag{7.3}$$

K_r depends on the surface coverage, S_r, by the restrainer where S_r is given by Equation (7.4):

$$S_r = C_{ads}/C_{sat} \tag{7.4}$$

Here, C_{ads} = restrainer adsorbed to surface, and C_{sat} = adsorption at saturation.

The magnitudes of C_{sat} and C_{ads} may not be well defined for compounds that chemically react with the surface (chemisorption). This might affect the absolute value of S_r. However, the following derivation would still be valid.

Two boundary conditions, which depend on the surface coverage by the restrainer, S_r, must be considered for the value of K_r. The first boundary condition exists when no restrainer is adsorbed to the surface ($S_r = 0$):

$$K_r = K_i \tag{7.5}$$

For the second boundary condition, at maximum surface coverage, the surface integration will be completely inhibited. Therefore, when $S_r = 1.0$, we obtain

$$K_r = 0 \tag{7.6}$$

Many functions may be proposed that may describe the dependence of K_r on the restrainer concentration and adsorption and that fulfill the boundary conditions of Equations (7.5) and (7.6). However, it appears reasonable to choose a simple model as a first approximation. Such a model is given by Equation (7.7):

$$K_r = K_i(1.0 - S_r) \tag{7.7}$$

It can be easily verified that this equation fulfills the boundary conditions set forth in Equations (7.5) and (7.6).

Substitution of Equation (7.7) into Equation (7.3) leads to Equation (7.8), which relates the nucleation to the surface coverage by restrainer:

$$Zr = K/[K_i(1.0 - S_r)] \tag{7.8}$$

This equation predicts that with increasing restrainer adsorption, i.e., increasing S_r, the product of Zr increases.

The model will now be expanded to relate the surface coverage, S_r, to the restrainer concentration.

Surface Adsorption Model

Many cases of adsorption to silver halides can be described by Langmuir adsorption isotherms. This equation describes the adsorption of many dyes and other compounds and has also been shown to be valid for the adsorption of the restrainer adenine to silver halides.[2] Thus, the Langmuir equation (Equation [7.9]) was used to obtain an estimate of S_r. If different adsorption mechanisms have been determined for a specific case, other adsorption equations may be substituted:

$$S_r = K_L C_L / (K_L C_L + 1.0) \tag{7.9}$$

Here, K_L is the Langmuir constant for adsorption and C_L is the restrainer concentration in solution in equilibrium with the adsorbed restrainer. K_L is a function of the adsorbent, substrate, solution composition, and temperature.

Since the silver halide surface charge depends on the pAg, ($-\log [Ag^+]$) ($[Ag^+]$ is the silver ion activity), and since this may affect the adsorption, K_L may be pAg dependent. The same is the case if the adsorbent reacts with Ag^+.

In cases where the adsorbent can react with hydrogen ions, K_L will also be pH dependent. Since H^+ and Ag^+ effects must be considered independently for each adsorbent, we will assume constant H^+ and Ag^+ concentrations for the continued discussion of the restrainer effect on nucleation.

For insertion into Equation (7.8), Equation (7.9) is transformed into Equation (7.10):

$$(1.0 - S_r) = 1.0/(1.0 + K_L C_L) \tag{7.10}$$

Insertion of Equation (7.10) into Equation (7.8) leads to Equation (7.11):

$$Zr = K(1.0 + K_L C_L)/K_i \tag{7.11}$$

Equation (7.11) predicts that a significant restrainer effect will only be observed when $K_L C_L > 0$.

Rewriting the equation and substituting Equation (7.2), we obtain Equation (7.12) where Z_0 and r_0 are the crystal number and size in the absence of growth restrainer:

$$Zr = Z_0 r_0 + (KK_L/K_i)C_L \tag{7.12}$$

The factors in the bracket are constants under given precipitation conditions. Thus, Equation (7.12) predicts that the product Zr should increase as the restrainer concentration, C_L, increases.

Since C_L, by definition, is equal to the solution concentration in equilibrium with adsorbed restrainer at the surface, it is difficult to give a precise value of C_L throughout the nucleation stage. Before nucleation, C_L is equal to the total restrainer concentration, C_{tot}. In the course of nucleation and precipitation, part of the original adsorbent attaches to the newly formed crystal surface. This reduces C_L from the original value of C_{tot} in a way that presently cannot quantitatively be predicted.

For further discussion, we will assume that C_L is not significantly different from C_{tot} during the nucleation phase precipitation.

The theoretical derivation thus far has related the change of Zr to the restrainer concentration. However, after the initial stages of kinetically controlled nucleation and crystal growth, diffusion-controlled growth dominates the growth processes. At this time, the number of crystals becomes constant, and Z no longer decreases with increasing crystal size. This was experimentally shown in Chapter 4.

At this stage, the constant number of crystals, Z_c, can be substituted for Zr in Equation (7.12). Together with the conclusions of the discussion relating C_L to C_{tot}, we obtain Equation (7.13), which relates Z to C_{tot}:

$$Z_c = Z_0 + f(KK_L/K_i)C_{tot} \tag{7.13}$$

The factor, f, takes into account the change from kinetically controlled to diffusion-controlled growth mechanisms, i.e., the change from the dependence on Zr to $Z_c =$ const. The predictions of Equation (7.13) will be experimentally tested.

EXPERIMENTAL

The compound 1-(3-acetamidophenyl)-mercaptotetrazole was used as restrainer (Table 7.1).[7] Silver chloride was formed by double-jet precipitation without and in the presence of restrainer.

TABLE 7.1
Molecular Structure of 1-(3-acetamidophenyl)-mercaptotetrazole

The restrainer was added to the gelatin solution, and the temperature and pAg were adjusted to the desired values. The initial addition rate was 0.0032 mole Ag/min. After 4 minutes, the addition rate was linearly accelerated over 20 min to 0.048 mole AgNO₃/min. This led to larger crystal sizes than nonaccelerated precipitations of equal time. The resultant larger crystal sizes allowed more accurate crystal size determinations.

The crystal size was determined by analysis of electron micrographs. For cubic crystals, the sizes are reported as cubic edge lengths. At high restrainer concentrations, the crystals were polymorphic, which was approximated by spherical morphology for the calculations. Crystal numbers, Z, were calculated from the average crystal volume and the volume of precipitated silver halide.

The crystal sizes, numbers of crystals formed, and crystal morphology are listed in Table 7.2.

RESULTS

The results in Table 7.2 show that the number of crystals increases as the restrainer concentration increases. Above 100 mg/l, significant changes in crystal morphology

TABLE 7.2
Silver Chloride Precipitation without and in the Presence of Restrainer

No.	Restrainer[a] mg/l	Crystal size[b] (μm)	$Z^c \times 10^{15}$	Morphology
1	0	0.734	0.035	Cubic
2	1	0.537	0.088	Cubic
3	3	0.465	0.137	Cubic
4	6	0.419	0.187	Cubic
5	10	0.332	0.374	Cubic
6	30	0.261	0.775	Cubic
7	100	0.211	2.780	Spheres
8	300	0.177	4.730	Spheres
9	500	0.135	10.700	Spheres

[a] 2(3-acetamidophenyl)-mercaptotetrazole; [b] Cubic: edge length, spheres: diameter; [c] Crystal number

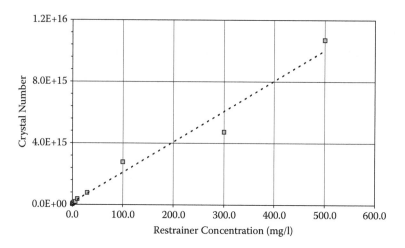

FIGURE 7.4   Crystal number versus restrainer concentration (mg/L).

are observed as shown in Figure 7.1. The crystal morphology under these conditions is relatively undefined, and thus, for the calculation of the crystal number, spherical morphology was used.

This results in some underestimation of the crystal number. In addition to a change in crystal morphology, a significantly larger, mono-modal crystal size distribution is observed (Figure 7.1). As discussed above, these show that the restrainer adsorbs to the crystal surface and alters the crystal growth processes.

The decrease in crystal size and increase in crystal number is in agreement with the predictions of Equations (7.12) and (7.13). In Figure 7.4, the crystal number is plotted versus restrainer concentration as suggested by Equation (7.13). A linear plot is obtained in agreement with the prediction. Linear least-squares analysis of the data leads to Equation (7.14):

$$Z = (1.91 \pm 0.12) \times 10^{13} C_{tot} + (10.2 \pm 23.5) \times 10^{13}$$

$$\text{Correlation coefficient} = 0.974 \tag{7.14}$$

The errors in the equation represent one standard deviation. The analytical correlation is in good agreement with the theoretical results of Equation (7.13): The number of crystals should increase linearly with restrainer concentration.

The linear correlation of Figure 7.4 limits the plot of the full range of data that extends over almost three orders of magnitude. The data are thus replotted in Figure 7.5 as a double-logarithmic plot.

Here, the linear correlation is shown over the full range of restrainer concentration and crystal numbers. The plot is linear over the full range of concentrations. The crystal number of the precipitation without restrainer is indicated by the line parallel to the concentration axis. The effectiveness of the restrainer is indicated by the

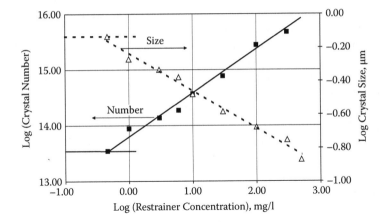

FIGURE 7.5   Log (crystal number) and (crystal size) versus log (restrainer concentration).

substantially increased crystal number by as low a level as 1.0 mg/l. The analytical evaluation of the log/log plot in Figure 7.5 leads to Equation (7.15):

$$\log Z = (0.789 \pm 0.0423)\log C_{tot} + (13.78 \pm 0.069)$$

$$\text{Correlation coefficient} = 0.983$$

(7.15)

Since all other precipitation conditions were held constant, a linear logarithmic correlation is obtained for size versus restrainer concentration. This plot is included in Figure 7.5. As expected, this correlation is linear. The linear equation is given in Equation (7.16).

$$\log(size) = -0.2248 * \log C_{tot} - 0.2332$$

$$\text{Correlation Coefficient} = 0.9872$$

(7.16)

A slope of 1.0 was predicted for Equation (7.15). The experimentally observed lower slope of 0.789 may be the result of the approximation of spherical morphology for the polymorphic crystals at higher restrainer levels. This approximation overestimates the crystal volume and thus underestimates the crystal number, resulting in the experimentally observed slope of less than 1.

CONCLUSIONS

The use of restrainers provides a powerful technique to reduce crystal size, "nanosizing" without changing other precipitation variables. Quantitative correlations between crystal size and number and restrainer concentration provide efficient size control.

The effect of growth restrainers on the nucleation of crystals was derived from the general BNG theory of crystal nucleation. The effect is based on adsorption of the restrainer to the crystal surface during the nucleation phase and a proportional decrease in the rate of surface integration of reactant material.

The model predicts that the crystal number increases with restrainer concentration present during the nucleation phase of precipitation. This was confirmed for precipitation of silver chloride in the presence of 1-(3-acetamidophenyl)-mercaptotetrazole over a range from 1 to 500 mg/l.

The present model of the restrainer effect on crystal nucleation should be generally applicable to other precipitation systems. The growth-reduction of critical nuclei by growth restrainers during the nucleation phase is a mechanism for the size control of nano particles.

REFERENCES

1. Leubner, I. H. 1987. Crystal formation (nucleation) in the presence of growth restrainers. *J Crystal Growth* 84:496–502.
2. James, T. H., ed. 1977. *The theory of the photographic process.* 4th ed. New York: MacMillan.
3. Szucs, M. 1978. *J Signal AM* 6:381.
4. Ammon-Brass, H., ed. 1971. *Restrainers in photographic gelatin.* Fribourg, Switzerland.
5. Strong R. W., and J. S. Wey. 1977. *Photogr Sci Eng* 21:14.
6. Strong, R. W., and J. S. Wey. 1979. *Photogr Sci Eng* 23:344.
7. Maskasky, J. E. 1986. *J Imaging Sci* 30:247.

8 Crystal Growth and Renucleation

CRYSTAL GROWTH, RENUCLEATION, AND MAXIMUM GROWTH RATE: THEORY AND EXPERIMENTS

INTRODUCTION

Renucleation is the formation of new crystals when the reactant addition rate exceeds the maximum growth rate of a growing crystal ("seed") population. This reaction occurred as an unwanted result during growth of crystals where it was desired to grow a shell on a seed population. The desire to control this process triggered the current treatment and modeling.[1]

To address the experimental need to avoid renucleation, a quantitative theory was derived. It is based on the mass-balance model for crystal growth and on the BNG theory presented in previous chapters.

The renucleation model relates the number of renucleated crystals formed, Z_n, to the reactant addition rate and to the seed crystal number, Z_s, or total seed amount, M_s, the seed surface area, and the seed crystal morphology.

The theory predicts that Z_n increases linearly with addition rate and decreases linearly with the amount or number of the seed crystals when renucleation occurs. The limiting reaction conditions where renucleation ends and where renucleation does not occur are also quantitatively given.

From the limiting condition for renucleation, the maximum growth rate of the seed crystals, G_m, was determined. The theory was experimentally verified by renucleation experiments using an octahedral silver halide dispersion in gelatin ("emulsion") having a silver bromide surface. The experimental growth rate of $11.4 < 12.4 < 13.6$ (A/sec, error limit is one standard deviation) was found in reasonable agreement with a value of 15.3 reported for similar precipitation conditions.

The precise determination of the maximum growth rate provides a fundamental variable of the BNG model.

The transition from crystal growth to renucleation has been used qualitatively to determine the maximum growth rate of crystals.[2,3,4] It is the object of this study to present and experimentally support a quantitative theory of renucleation.

Uncontrolled renucleation is undesirable because it leads to irreproducible crystal populations. An example of renucleation is shown in Figures 8.1 and 8.2, where crystal pictures and crystal size distributions of the seed emulsion and emulsions with growth only (no renucleation) and with growth and renucleation are shown.

In view of the technical importance of controlling renucleation and its use to determine maximum growth rates, it is surprising that there existed no quantitative

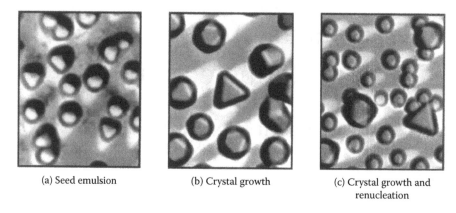

(a) Seed emulsion   (b) Crystal growth   (c) Crystal growth and renucleation

FIGURE 8.1   Electron microscope images for control, growth, and renucleation.

theory that correlated the onset of renucleation, the maximum growth rate, and the number of crystals formed in the renucleation step with the growth and renucleation variables and the properties of the seed emulsions. This chapter presents a theory that quantitatively relates the number of renucleated crystals in seeded emulsion precipitations to the reactant addition rate, the precipitation conditions, and

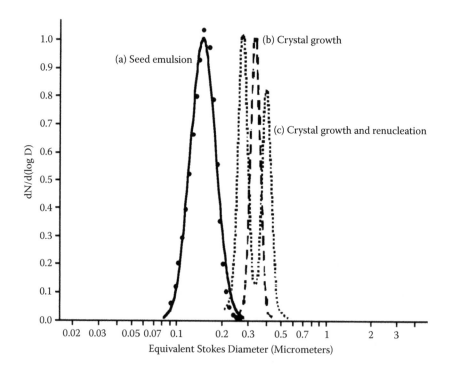

FIGURE 8.2   Crystal size distribution for control, growth, and renucleation.

the concentration and properties of the seed emulsion. The theory also allows a determination of the maximum growth rate, G_m, of the seeds under the experimental precipitation conditions and the nucleation rate without seed emulsion. This theory is based on the theory for crystal growth and on the nucleation theory derived by the author and coworkers.

GENERAL EXPERIMENTAL PROCEDURES

Experimentally, the present study follows previous work in which deliberate renucleation was used. The results were applied to determine the maximum growth rate and growth mechanism of silver chloride and silver bromide.[2–8] The experimental details for the present experiments are given in the experimental section after the derivation of the theory. Two types of experiments were used.

Control of Reactant Addition Rate: Fixed Seed Concentration

In a balanced double-jet precipitation, $AgNO_3$ and alkali halide solutions are added at a constant rate to an aqueous gelatin solution of silver halide ("seed") crystals. In a series of experiments, the addition rate is varied, while the seed emulsion concentration is held constant. Renucleation is detected by the presence of new small crystals at the end of the precipitation (Figure 8.1). In previous studies, the critical addition rate at which renucleation begins was derived by extrapolating between addition rates where the transition to renucleation occurred.

Control of Crystal Seed Concentration: Fixed Addition Rate

This is the most effective approach to determine the maximum growth rate. It was suggested by the theoretical derivation and was reported for the first time.[1] In these experiments, the addition rate was constant throughout the experiments, and the concentration of the seed emulsion was varied. The critical growth rate was determined from the seed concentration where renucleation ends.

Although these and previous procedures should give the same results for the critical growth rate, significant differences were reported.[2] This may be due to a crowding effect on the crystal growth rate at high suspension densities.[2–8] In addition, the precision and accuracy of the results depends on the method of interpolation used to determine the onset or end of renucleation.

THE RENUCLEATION MODEL

The model is based on these fundamental concepts:

1. During renucleation the seed crystals grow at maximum growth rate.
2. Material added in excess of the maximum growth rate of the system causes renucleation.
3. The mechanism of nucleation with seeds (renucleation) is the same as without seeds and is given by the BNG theory.

Crystal Nucleation

The number of crystals formed under diffusion-controlled growth conditions, Z, is given by Equation (8.1), since the growth of silver chloride, bromide, and iodide is determined by diffusion-limited processes. For kinetically controlled growth, Equation (8.1) must be substituted by the nucleation equation for kinetic conditions, as derived earlier. However, for most renucleation conditions in silver halide precipitations, the present derivation should apply.

$$Z = \left[\frac{R_g}{2k_v \gamma D V_m} \right] \frac{RT}{C_s(r/r^* - 1.0)} \tag{8.1}$$

Here, Z is the number of crystals formed, R the molar addition rate, C_s the solubility, T the temperature (K), R_g the gas constant, and D the diffusion constant. The following are properties of the crystal population: the surface energy, γ, the crystal molar volume, V_m, the average crystal size, r, and the critical crystal size, r*. The critical crystal size, r*, is the size at which a crystal has equal probability to grow or to dissolve by Ostwald ripening. The size, r*, represents a boundary condition and not a crystal population.

CRYSTAL GROWTH

At and below the maximum growth rate, the crystal number, size, and total mass are given by the mass balance (Equation [8.2]). This mass-balance equation is not valid when the (nominal) growth rate exceeds the maximum growth rate, G_m.

$$Zk_v L^3 = MV_m \tag{8.2}$$

Here, Z is the number of crystals, k_v is the crystal volume factor (1.0 for cubic crystals, 0.4714 for octahedra, where the size, L, is given by the edge length), M is the number of moles of crystals, and V_m is the molar volume.

Growth Rate and Crystal Number

The crystal growth rate is derived from the mass balance (Equation [8.2]) when reactants are added to grow the crystal population (Equation [8.3]).

$$G = \frac{dL}{dt} = \frac{R_s V_m}{3.0 k_v L^2 Z_s} \tag{8.3}$$

Here, G is the crystal growth rate, which is defined by the change in size per time unit, dL/dt; R_s is the fraction molar addition rate (dM/dt), which contributes to the growth of the seed crystals; V_m is the molar volume; and Z_s is the number of seed crystals in the reactor.

Growth Rate and Surface Area

At times, it is more convenient to use the total surface area of the seed crystal population than their size to determine their growth rate. This may be a practical solution if the seed population is polymorph or has a diversity of surface areas. The surface area, S, of the seed crystal population, Z_s, is given by Equations (8.4) and (8.5).

$$S = k_s L^2 Z_s \qquad (8.4)$$

$$S = M_s S_m \qquad (8.5)$$

Two additional equations, (8.6) and (8.7), can be derived from Equations (8.3)–(8.5) to give equivalent expressions for the growth rate that may be used under specific conditions.

$$G = \frac{R_s k_s V_m}{3.0 k_v M_s S_m} \qquad (8.6)$$

In Equation (8.6), M_s is the moles of seed emulsion added to the reactor, S_m the surface area/mol of seed crystals, and k_s the crystal surface shape factor (6.0 for cubes, 3.4641 for octahedra, when the size is given as edge length). The other variables and constants were given previously.

The molar surface area may be determined by adsorption methods, e.g., with dyes, which are independent of a knowledge of the crystal size and morphology. In cases in which k_s and k_v are not known, for instance, when the morphology of the crystal population is mixed or unknown, an apparent growth rate may be determined that includes these parameters.

Growth Rate and Crystal Size

If the mass-balance equation, (8.2), is solved for Z, and inserted in the basic growth equation, (8.3), Equation (8.7) is obtained:

$$G = \frac{R_s L_s}{3.0 M_s} \qquad (8.7)$$

Equation (8.7) relates the growth rate, G, to the addition rate, R_s, the crystal size of the seed emulsion, L_s, and the molar amount of seed crystals, M_s. When G is determined by these equations, its magnitude includes the size definition of the crystal size measurement. G will be different if the size of a crystal is determined by edge length, diameter, radius, or some other characteristic size.

During renucleation, the original crystals grow at maximum growth rate, and the growth rate, G, in Equations (8.3), (8.6), and (8.7) must be replaced with the maximum growth rate, G_m. It is further postulated that the number of crystals formed during renucleation is given by Equation (8.1).

Substitutions

To simplify the further development of the model, the substitutions a, b, and c are introduced.

From Equation (8.1):

$$a = \frac{R_g T}{2k_v \gamma V_m DC_s (r/r^* - 1.0)} \tag{8.8}$$

From Equation (8.3):

$$b = \frac{3.0 k_v L^2 G_m}{V_m} \tag{8.9}$$

From Equations (8.6) and (8.7) two expressions for a constant c can be derived (Equations [8.10] and [8.11]).

$$c = \frac{3.0 k_v S_m G_m}{k_s V_m} \tag{8.10}$$

$$c = \frac{3.0 G_m}{L_s} \tag{8.11}$$

GROWTH AND RENUCLEATION

Under renucleation conditions, the mass balance demands that the total addition rate, R, be equal to the sum of the addition rate fractions, R_n and R_s, which are consumed by renucleation and growth, respectively (Equation [8.12]):

$$R = R_n + R_s \tag{8.12}$$

In precipitations where a significant amount of material is in solution, an additional term must be added for the material that stays in solution. For highly insoluble materials such as those considered herein, this factor is insignificantly small compared with the material consumed for growth and renucleation.

Solving Equation (8.1) for R_n leads to Equation (8.13). To solve for R_s, Equations (8.3), (8.6), (8.7), and (8.8)–(8.11) are combined, leading to Equations (8.14) and (8.15).

$$R_n = Z_n / a \tag{8.13}$$

$$R_s = b Z_s \tag{8.14}$$

$$R_s = c M_s \tag{8.15}$$

In the next step, Equations (8.13)–(8.15) are substituted into Equation (8.12), which is solved for the number of recrystallization crystals, Z_s, which leads to Equations (8.16) and (8.17).

$$Z_n = aR - abZ_s \qquad\qquad (8.16)$$

$$Z_n = aR - acM_s \qquad\qquad (8.17)$$

Equations (8.16) and (8.17) are equivalent in describing renucleation. In Equation (8.16) the number of seed crystals, Z_s, and the constant b represent the effect of the seed population; in Equation (8.17) the molar amount of seed crystals, M_s, and the constant c do the same. In Equation (8.17), the constant c contains the molar surface area, S_m, or the crystal size, L_s (Equations [8.10] and [8.11]), and it can thus readily be rewritten for seed emulsions of different surface area or size. As can be seen from Equations (8.9)–(8.11), constants b and c also contain the maximum growth rate. Thus, G_m can be calculated once b and c have been experimentally determined.

The maximum growth rate, G_m, is also a function of temperature, solubility, pAg, crystal morphology, crystal composition, growth mechanism, and other features modifying the crystal growth processes.[2–7]

ASPECTS OF THE RENUCLEATION MODEL

Determination of a, b, and c

Equations (8.16) and (8.17) can take on positive and negative values and the value of zero. Negative and zero values indicate that only growth and no renucleation occurs. The magnitude of the negative values may be considered as a measurement related to a resistance to renucleation. The equations thus reflect the experimental observation that, depending on reaction conditions, renucleation may or may not be observed.

To describe renucleation quantitatively, constants a, b, and c in Equations (8.16) and (8.17) must be determined. The values of a, b, and c can be determined only when Z_n is positive, that is, when renucleation occurs. Conditions where no renucleation is observed may be either equal to or below the maximum growth condition ($Z_n < 0.0$).

The values of a, b, and c may be determined from Equations (8.16) and (8.17) by varying the addition rates and seed concentrations randomly in statistically designed experiments, and determining a, b, and/or c from the experiments where renucleation is observed.

However, for the confirmation of the theoretical model, it was decided to choose experimental conditions where either the addition rate or the seed concentrations were varied independently.

Variable Seed Concentration

The seed concentration (mole) is varied and the addition rate is held constant. A linear correlation of Z_n versus M_s with a negative slope is predicted. The intercept with the Z_n axis ($M_s = 0$) represents nucleation without seeds and is equal to aR. The

TABLE 8.1

Z_n as a Function of Seed Crystal Concentration and Number

	Seed Emulsion	Size (µm)		$Z \times 10^{12}$			Inter-
No.	M_s (mol)	Seed L_s	Renucleation,L_n	Z_s	Z_n	R/M_s	Distance r_i
1	0.0000		0.191	0	6111		
2	0.0075	0.362	0.218	158	3478	3.98	21.0
3	0.0113	0.366	0.237	237	2404	2.65	18.3
4	0.0151	0.403	0.271	316	1165	1.99	16.6
5	0.0189	0.419	0.320	394	459	1.59	15.4
6	0.0226	0.429	0.348	473	168	1.33	14.5
7	0.0302	0.407		631	0	0.99	13.2
8	0.0603	0.337		1262	0	0.50	10.5
9	0.1207	0.274		2524	0	0.25	8.3

Constant addition rate of 0.030 mol/min. Flow rate: 10 mL/min, 3.0 N $AgNO_3$ and NaBr. Seed emulsion: 0.1506 µm (ecd). Size = equivalent circular diameter, disc centrifuge. Z_s = number of seed crystals. Z_s = number of renucleation crystals. R/M_s = ratio of addition rate to moles of seed emulsion. F_1 = relative interparticle distance (Ref. 5).

intercept with the M_s axis gives the critical renucleation condition where the seed crystals are at maximum growth rate. Above this seed concentration, renucleation is not observed (Table 8.1). The discussion is the same when M_s is replaced with the seed number, Z_s.

Variable Addition Rate

An alternate approach for determining a, b, and c is keeping the seed population (expressed as Z_s or M_s) constant and varying the addition rate, R. As long as renucleation is observed, a linear correlation of Z_n versus R with a positive slope of "a" is expected. The intercept with the R-axis gives the critical renucleation condition where the seed crystals are at maximum growth rate. Below this addition rate, renucleation is not observed (Table 8.2).

The experiments where both R and Z_s or R and M_s are varied can be combined, and least-squares linear correlations that give constants a, b, and c, as well as an intercept, can be calculated. A significant value for the intercept indicates a significant difference between changing the addition rate and the seed crystal population for the renucleation experiments. The experimental results will show no significant difference.

No Seed Crystals

If no seed crystals are present ($M_s = Z_s = 0$), Equations (8.16) and (8.17) revert to Equation (8.1). This means that nucleation without seeds can be considered a limiting case of renucleation with seeds.

TABLE 8.2
Z_n as a Function of Addition Rate

No.	Addition rate (mol/min)	Size (µm)		$Z \times 10^{12}$		R/M$_s$
		Seed	Renucleation	Z_s	Z_n	
1	0.015	0.303	0	631	0	0.50
2	0.030	0.407	0	631	0	0.99
3	0.045	0.429	0.330	631	464	1.49
4	0.053	0.408	0.286	631	1462	1.74
5	0.060	0.394	0.266	631	2637	1.99
6	0.068	0.383	0.247	631	4177	2.24
7	0.075	0.378	0.233	631	5946	2.49
8	0.083	0.375	0.225	631	7565	2.73
9	0.090	0.374	0.201	631	7972	2.98

Constant seed amount: 0.0302 mol (20.0 g), $r_i = 13.2$. Seed emulsion: 0.1506 µm (ecd); Size = µm, equivalent circular diameter, disc centrifuge. Z_s = number of seed crystals; Z_n = number of renucleation crystals. R/M$_s$ = ratio of addition rate to moles of seed emulsion.

Maximum Growth Rate Conditions

Under the experimental conditions where renucleation ends ($Z_n = 0$), Equations (8.16) and (8.17) simplify to Equations (8.18) and (8.19).

$$R/Z_s = b \qquad (8.18)$$

$$R/M_s = c \qquad (8.19)$$

Thus, b and c can be calculated from the limiting renucleation conditions.

Renucleation Conditions

Renucleation is obtained when $Z_n > 0$. This condition is given when Equations (8.20) and (8.21) are fulfilled.

$$R/Z_s > b \qquad (8.20)$$

$$R/M_s > c \qquad (8.21)$$

Thus, the renucleation conditions are uniquely defined by b and c. Once b and/or c have been determined, it is possible to predict the renucleation conditions as a function of the ratio of addition rate to number of seed crystals (constant b, Equation ([8.20]) or the ratio of addition rate to the molar amount of seed crystals (constant c, Equation [8.21]).

EXPERIMENTAL

Seed Emulsion

The seed emulsion was an octahedral AgBrI emulsion that was nucleated and grown to 50% of its volume with 10/90 mol% sodium iodide/bromide, and the outer shell (50% of volume) grown with sodium bromide. Thus, for the renucleation experiment, this emulsion was considered to have an AgBr surface. The size was determined by Joyce-Loebl disc centrifuge to be 0.1506 µm equivalent circular diameter (ecd), which corresponds to an octahedral edge length of 0.1434 µm (0.952 × ecd). The molar surface area of this emulsion was experimentally determined by adsorption of 1,1'-diethyl-3,3'-cyanine dye to be 1.256 m²/mol.

It is known that the particle density can affect the maximum crystal growth rate.[8] The relative interparticle distance, r_i, was calculated. For the seed-emulsion concentration range where renucleation occurs (Tables 8.1 and 8.2), the value of r_i is > 13.2. This value is in the range of noncrowded conditions where the maximum growth rate is independent of the seed-emulsion concentration.[7]

Precipitation Procedure for Growth/Renucleation Experiments

Deionized bone gel (80 g) was dissolved in 1,900 g distilled water at 40°C. Seed emulsion was added as indicated in Tables 8.1 and 8.2. Then the pH was adjusted to 5.6 at 40°C. The temperature was raised to 70°C, and the pAg was adjusted to 8.50 with 3.0 N NaBr. Then AgBr was precipitated for 20 min at pAg 8.50 with constant flow rates (mL/min), as indicated in Tables 8.1 and 8.2 (3.0 N AgNO₃ and NaBr). At the end of the precipitation, 120 g of emulsion was split off, and 2.0 mL of a solution of 50 mg/mL 4-hydroxy-6-methyl-1,3,3a-tetraazaindene in water was added to retard Ostwald ripening. For low-emulsion concentrations, the original emulsion samples were concentrated by centrifuging and redispersion in water.

The crystal sizes were determined by electron image analysis (EIA) and Joyce-Loebl disc centrifuge.[3] The limit of error for the crystal sizes is ± 5%. Electron micrographs using shaded carbon replicas were also obtained (Figure 8.1).

DETERMINATION OF THE NUMBER OF RENUCLEATION CRYSTALS, Z_N

The disk-centrifuge and EIA size determinations give relative values of the crystal populations that may be used for the calculations. However, at low-emulsion concentrations, the original emulsion sample must be concentrated by centrifuging, and part of the crystal populations may be lost. For this reason, and because of the feeling that the determination of crystal size populations is generally relatively inaccurate, this method was not used. Thus, the calculation of the relative crystal populations from the crystal sizes was preferred.

The following information is available from the experimental conditions, the information on the seed emulsion, and from analytical data:

- Seed emulsion: moles added, M_s; initial crystal size, L_s; final crystal size, L_f
- Added silver halide: M_t
- Renucleated population: final size, L_n

The total molar amount of the seed emulsion after precipitation, M_f, can be calculated by using Equation (8.22):

$$M_f = M_s (L_f / L_s)^3 \tag{8.22}$$

From the molar mass balance, the molar amount of the renucleation population, M_n, is obtained from Equation (8.23):

$$M_n = M_t + M_s - M_f \tag{8.23}$$

The number of renucleated crystals, Z_n, can then be determined with Equation (8.24):

$$Z_n = \frac{M_n V_m}{k_v L_n^3} \tag{8.24}$$

The values of Z_n are included in Tables 8.1 and 8.2. The accuracy of the Z_n values is mainly determined by the accuracy of crystal size determinations ($\pm 5\%$) and the effect of crystal morphology, as expressed by k_v.

RESULTS AND DISCUSSION

The experimental conditions and results are summarized in Tables 8.1 and 8.2. For the evaluation of the data, the addition rate is given in units of mol/min, the amount of seed emulsion in moles, and the number of seed crystals in units of 10^{12}.

In Table 8.1, the amount of seed emulsion was varied from zero to 0.1207 mole, and the addition rate was held constant at 0.030 mol Ag/min. The data show that the number of renucleated crystals decreases with increasing amount of seed emulsion. Renucleation ends between 0.0226 and 0.0302 mol seed emulsion. The ratio of addition rate and moles of seed emulsion, R/M_s, indicates that the critical ratio where renucleation ends is between 0.99 and 1.33.

In Table 8.2, the amount of seed emulsion was held constant at 0.0302 mol, and the addition rate was varied from 0.015 to 0.090 mol Ag/min. Renucleation began between 0.030 and 0.045 mol/min, at R/M_s ratios between 0.99 and 1.49.

In Figures 8.3 to 8.5, the experimental results (Tables 8.1 and 8.2) are plotted according to Equations (8.16) and (8.17) as Z_n versus Z_s, Z_n versus M_s, and Z_n versus R, respectively.

As predicted by the equations, linear correlations are obtained when the seed crystal number (Figure 8.3) or amount (Figure 8.4) fall below a critical limit, or when the addition rate exceeds a limiting value (Figure 8.5). For the ranges where Z_n depends on the precipitation conditions, the results were quantitatively evaluated according to Equations (8.16) and (8.17).

From the experimental data in Table 8.1, where the seed concentration was varied, Equations (8.25) and (8.26) are obtained for Z_n versus Z_s and Z_n versus M_s.

$$Z_n = (5.94 \pm 0.24) 10^{15} - (14.56 \pm 0.76) 10^{12} Z_s \tag{8.25}$$

$$Z_n = (5.94 \pm 0.24) 10^{15} - (304.7 \pm 16.2) 10^{15} M_s \tag{8.26}$$

$$R^2 = 0.992$$

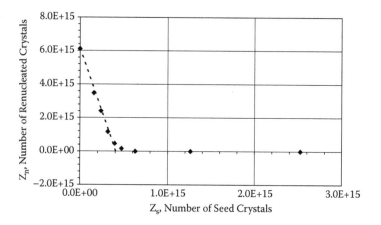

FIGURE 8.3   Number of renucleated crystals Z_n as a function of the number of seed crystals Z_s.

Here M_s is given in moles. The limits of error in the equations represent one standard deviation. The correlation coefficient is 0.992, which confirms the linear correlations observed in Figures 8.3 and 8.4.

The data from Table 8.2 give the Z_n versus R correlation in Equation (8.27).

$$Z_n = (181.1 + 9.6)10^{15}R - (7.91 \pm 0.38)10^{15} \qquad (8.27)$$

Here, R is given in mol/min. When these equations are extrapolated to the boundary conditions where renucleation begins ($Z_s = 0$), the following critical values are obtained for the present precipitation conditions: $Z_s = 4.08 \times 10^{17}$, $M_s = 0.0195$ (mol), $R = 43.7 \times 10^{-3}$ mol/min.

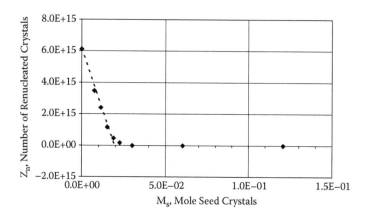

FIGURE 8.4   Number of renucleated crystals Z_s as a function of molar amount of seed crystals M_s.

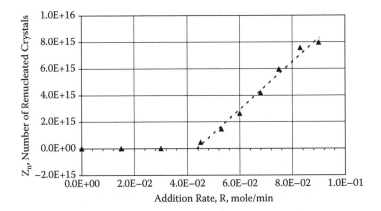

FIGURE 8.5 Number of renucleated crystals Z_n as a function of reactant addition rate.

The data in Tables 8.1 and 8.2 were combined and evaluated by linear regression for the correlation of Z_s with R and Z_s (Equation [8.28]).

$$Z_n = (186.9 + 7.2)10^{15}R(13.51 \pm 0.71)10^{12}\ Z_s + (149 \pm 254)10^{12}$$

$$R^2 = 0.987$$

(8.28)

Here, Z is given in units of 10^{12}. The intercept is within the limits of error equal to zero. This indicates that there is no interaction between the addition rate and the number of seed crystals. It also indicates that the critical growth rate is the same if it is determined by variation of addition rate or by seed concentration.

The correlation of Z_n with R and M_s is given by Equation (8.29).

$$Z_n = (186.9 + 7.2)10^{15}R - (281.8 \pm 14.9)10^{15}M_s + (141 \pm 256)10^{15}$$

$$R^2 = 0.987$$

(8.29)

The correlation coefficient and intercept are the same as for Equation (8.28). Again, the intercept indicates that there is no interaction between R and M.

Equations (8.28) and (8.29) can now be evaluated for the values of a, b, and c in Equations (8.16) and (8.17), resulting in a = 186.9, b = 72.29 × 10⁶, c = 1.51.

Equations (8.20) and (8.21) predict that renucleation occurs when $R/Z_s > b$ or $R/M_s > c$.

From Equations (8.9)–(8.11), the maximum growth rate, G_m, was calculated. The constants used in the equations are k_v (oct.) = 0.4714, k_s (oct.) = 3.46, $V_m = 29.0 \times 10^{12}$ µm³/mol, and octahedral edge length = 0.143 µm (0.1506 µm × 0.952).

The difference of the V_m values for AgBr (29.0 cm³/M AgBr) and AgBrI (5% I, 29.46 cm³, seed emulsion) is very small (1.6%), and the value for AgBr was used for all calculations. The surface area was determined to be 1.256 (m²/mol Ag). The values (and equations used) for the maximum growth rates (G_m, A/sec) are within one standard deviation of b and c (Table 8.3).

TABLE 8.3

Calculated G_m Values

Equation (8.9): $G_m = 11.1 < 12.2 < 13.4$ (A/sec)

Equation (8.10): $G_m = 12.0 < 13.1 < 14.3$ (A/sec)

Equation (8.11): $G_m = 11.0 < 12.0 < 13.2$ (A/sec)

Differences in the average growth rates arise from the errors in determining b and c, and from errors in the constants and experimental variables in Equations (8.9)–(8.11). One type of error results from the assumption that the crystals are perfect octahedra. Any deviation from this assumption is reflected in the values of k_v, k_s, and the octahedral edge length, L. These errors affect all three equations. Equation (8.10) further depends on the experimental error of the determination of the surface area, S_m.

From Jagannathan's data,[7] a growth rate of 15.3 A/sec was extrapolated if one corrects for k_v (0.4714, Equation [8.3]) of the octahedral emulsion (Ref. 7, 70°C, pBr 2.1, pAg 8.4). Considering the different experimental conditions and the different methods of extrapolation, the present data (range of 11.0–14.3 A/sec) are in good agreement with the value of Ref. 7 (15.3 A/sec).

SUMMARY AND CONCLUSIONS

A quantitative theory of renucleation was derived on the basis of the theory of crystal nucleation and the mass-balance model of crystal growth. Relatively simple equations were derived to predict the renucleation conditions. The predictions of the theory were experimentally confirmed for renucleation studies of octahedral AgBr crystals. The predictions of the renucleation theory were used to determine the limiting renucleation conditions and the maximum growth rate with improved precision. The number of renucleated crystals can be quantitatively related to the precipitation conditions.

The new model and theory greatly simplify the determination of the maximum growth rate and the renucleation conditions for any precipitation system. The model suggested a new efficient experimental procedure for renucleation and the determination of the maximum growth rate where nucleation conditions were held constant and the concentration of seed crystals was varied.

REFERENCES

1. Leubner, I. H. 1993. Crystal growth and renucleation: Theory and experiments. *J Imaging Sci Technol* 37:510–16.
2. Wey, J. S., and R. W. Strong. 1977. *Photogr Sci Eng* 21:14.
3. Wey, J. S., and R. W. Strong. 1977. *Photogr Sci Eng* 21:248.
4. Strong, R. W., and J. S. Wey. 1979. *Photogr Sci Eng* 23:344.
5. Jagannathan, R., and J. S. Wey. 1981. *J Cryst Growth* 51:601.

6. Jagannathan, R., and J. S. Wey. 1982. *Photogr Sci Eng* 26:61.
7. Wey, J. S., and R. Jagannathan. 1982. *AIChE J* 28:697.
8. Jagannathan, R. 1988. *J Imaging Sci* 32:100.
9. King, T. W., S. M. Shor, and D. A. Pitt. 1981. *Photogr Sci Eng* 25:70; technical details of size determination by disc-centrifuge should be obtained from the manufacturer.

9 Continuous Crystallization

CONTINUOUS CRYSTALLIZATION IN THE MIXED-SUSPENSION, MIXED-PRODUCT-REMOVAL (MSMPR), OR CONTINUOUS STIRRED TANK REACTOR (CSTR) CRYSTALLIZER

INTRODUCTION

Continuous production of materials is an important industrial process. For continuous chemical processes and crystallizations, the controlled continuous stirred tank reactor (CSTR), or mixed-suspension, mixed-product-removal (MSMPR) system, and the stop-flow system are preferred for controlled crystallizations. The present work concentrates on the CSTR (MSMPR) crystallizer and on the effect of varying crystal solubility on the crystal size. A schematic sketch of a CSTR reactor for precipitation of silver bromide is shown in Figure 9.1.

In crystallization processes and for the control of crystal size, growth, and design, the control of crystal nucleation is a crucial first step in research, product development, and manufacturing. For this purpose, the balanced nucleation and growth model (BNG) was developed.[1] The BNG model is the basis that provides a fundamental model to control crystal nucleation and growth in continuous crystallizations. Experimental results will be presented that support the model.

EXPERIMENTAL

The reactants, sodium bromide and silver nitrate solutions (Figure 9.1), are added at a controlled flow rate. A gelatin solution, which is adjusted to the aim pAg and is heated to aim temperature, is also added at a controlled flow rate. Gelatin concentration and dilution are adjusted to balance the loss of gelatin and water in the product flow. The product flow contains the silver halide, in this case silver bromide, sodium bromide, gelatin, and water. It also is at the aim reaction temperature and pAg.

The product yield is determined by the reactant addition rate and the residence time. The residence time, τ, is defined as the reaction volume in the reactor divided by the sum of input or product flow rates, respectively. The properties of the resultant crystal population are determined by the reaction conditions, solubility, temperature, reactant addition rate, presence of restrainers[2] or ripeners,[3] and the residence time. The latter has a similar function as the length of precipitation time in batch precipitations. In addition, it has been generally assumed that the steady-state suspension density, e.g., mole AgBr/l, in the reactor affects the crystal population properties.

It is the intent of the present and the following sections to evaluate and quantify the effect of these variables both theoretically and by controlled experiments.

In both batch and continuous crystallization processes and for the control of crystal size, growth, and design, the control of crystal nucleation is a crucial first step

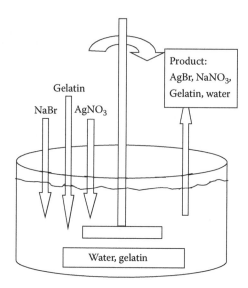

FIGURE 9.1 Continuous stirred tank reactor (CSTR).

in research, product development, and manufacturing. For this purpose, the balanced nucleation and growth (BNG) model was developed for batch processes.[14] We will show that the application of the concepts of the BNG model is the basis that provides a fundamental model to control crystal nucleation and growth in continuous crystallizations.

To obtain consistent crystalline products, it is important to understand the relationship between the engineering variables and the properties of the precipitate. Important crystallization engineering variables are reaction volume, temperature, crystal solubility, material flow and reactant addition rates, residence time, and the presence of ripeners and restrainers, suspension density, stirring/mixing, and the balanced withdrawal of reacted and unreacted material. The precipitate properties are determined by the composition of the precipitate, average crystal size, and crystal size distribution (CSD).

Related to the product properties are reactions taking place in the reactor, the formation of crystals (nucleation rate), their growth (growth rate), and the mechanisms of crystal formation, homogeneous, heterogeneous, or secondary nucleation.

PRIOR MODELS: THE RANDOLPH-LARSON MODEL

Different but complementary steady-state theories of crystallization in the continuous stirred tank reactor (CSTR, or mixed-suspension, mixed-product-removal, MSMPR) crystallizer have been proposed,[4–7] of which the Randolph-Larson model has been dominant.

Bransom, Dunning, and Millard and Randolph and Larson (R-L) modeled the mass-balance correlation between crystal size distribution, crystal growth rate, and residence time (Equation [9.1]) at steady state. While Bransom and colleagues gave the first derivation,[5] the model is rightfully named after Randolph and Larson who were the main proponents and contributors to its study and who published a summary of their work.[5,6]

$$n_x = n_0 \exp\left(-\frac{L_x}{\tau G}\right) \qquad (9.1)$$

In Equation (9.1), $n_x = Z_x/cm \times cm^3$, where n_x is number population density of crystals of size L_x per reaction volume; Z_x is the number of crystals of size L_x; and $n_0 = Z_0/cm \times cm^3$ ("cm" represents the crystal size, "cm^3" the normalization to reaction volume) is the number of crystals, Z_0, of size L_0 nucleated per time and volume. L_x is the crystal size of defined crystal size populations; $G = dL/dt =$ growth rate; and $\tau =$ residence time, determined by the reaction volume divided by material flow rate (input or output). The underlying model (Equation [9.1]) is based on continuous nucleation of infinitely small crystals followed by growth. The model parameters are the residence time, τ, and crystal growth rate, G.

The model (Equation [9.1]) does not relate crystal size and distribution to reaction variables such as temperature, crystal solubility, and suspension density. To overcome this restriction, additional concepts and equations were introduced.[7] These include birth and death rates of nuclei and crystals, which represent the formation and loss of particles in the reaction mixture. In this model, birth (B^0) and growth rates (G) of the crystals determine the stability of the continuous crystallizer. A supra-linear correlation between these rates has been proposed, as given in Equation (9.2).

$$B^0 \sim G^i \qquad (9.2)$$

The exponent, i, is a proposed kinetic order. It was theoretically concluded that for high-yield systems (e.g., silver halides), the crystallization process would be inherently stable if the kinetic order was less than 21. For silver halide systems, i was found to be approximately 4, and thus these systems should not be cycling at steady state.[6,9]

GROWTH

The growth rate, G, is a function of addition rate and can be expressed by the mass-balance equation, (9.3).

$$G = \frac{dL}{dt} = \frac{RV_m}{3.0k_v ZL^2} \qquad (9.3)$$

Here, L is the crystal size, R is the addition rate (mole/s), V_m is the molar volume (cm^3/mole), k_v is the volume conversion factor that converts L to the crystal volume, and Z is the number of crystals in the reaction mixture. For polydisperse reaction mixtures, Z is replaced by Z_x and L by L_x, and it is necessary to integrate over x. G is also a function of supersaturation, which frequently cannot be experimentally determined or used as control variable. For further details of the Randolph-Larson model, it is necessary to refer to the literature.[7]

Generally, n_x, n_0, L_x, and G (Equation [9.1]), birth and death rates, supersaturation, and other R-L variables are not known for continuous crystallization systems that are based on continuous nucleation and growth. Equation (9.1) is, however, well suited for the control of crystal growth and crystal size distribution of continuously seeded CSTR systems, in which the input of number and size of seeds (n_0, birth rate) and the reactant addition rate for growth can be quantitatively controlled. This will be presented in a subsequent section.[8]

EXAMPLE OF SIZE DISTRIBUTION IN CSTR PRECIPITATION

For controlled continuous crystallizations of silver halide in the CSTR system, a well-defined crystal size distribution is obtained at steady state (Figure 9.2). Replotting the size distribution using the guidance of Equation (9.1) did not give the predicted linear plot (Figure 9.3).

The measurement cut-off for crystal size distribution at the time of this experiment was about 0.08 µm, and thus the increase in crystal number with crystal size is within the experimental sizing range. The lower crystal number at small sizes is difficult to bring in alignment with the prediction of Equation (9.1), which predicts a steady decrease of crystal number with increasing crystal size.

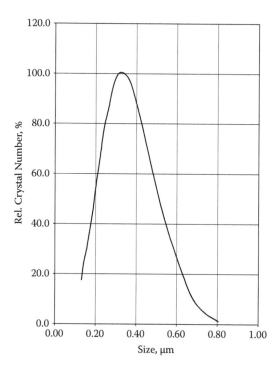

FIGURE 9.2   Size distribution of silver chloride (CSTR crystallizer).

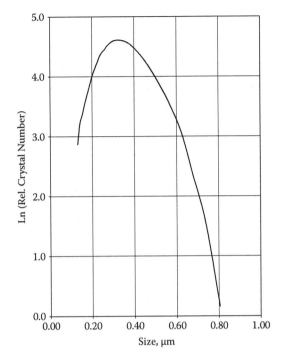

FIGURE 9.3 Randolph-Larson plot of AgCl (CSTR) size distribution.

An extensive study showed that the size distribution of continuous silver bromide crystallizations was not described by Equation (9.1). It was observed that only a small, large-sized part of the crystal population obeyed this equation.[9]

The practical control of crystallizations is achieved, as previously shown, by control of reactant addition rate, crystal solubility, temperature, resident time, and ripener and restrainer concentration. In our studies, we found that the Randolph-Larson model did not explicitly address these reaction variables and did not allow relating our experimental results to the practical control variables.

For many product applications, e.g., photographic systems, the most numerous ("maximum") crystal size population, L_m, dominates the critical product properties, e.g., photographic sensitivity. Thus, it became necessary to develop a model that correlates L_m with experimental control variables. The present BNG model predicts the maximum size of crystals obtained in controlled continuous crystallizations (CSTR, MSMPR) to the experimental control variables. The maximum crystal size is defined in this context as the size of the crystal population that has the highest number of crystals in the reaction mixture. For systems that do not deviate too much from monodispersity, this crystal size generally represents the general properties of the crystal size population.

In following sections, we will show experimental size correlation with residence time and crystal solubility, and relate them to the BNG model. We will show that the BNG model is free of adjustable parameters, e.g., the kinetic order, i, and provides many reaction parameters not available from the Randolph-Larson model. The test of the BNG model with precipitation temperature and ripener and restrainer concentration will have to wait for future experiments.

TRANSIENT BEHAVIOR OF SILVER BROMIDE PRECIPITATION IN A CONTINUOUS SUSPENSION CRYSTALLIZER

Summary

CSTR systems reach mass balance at about four residence times after startup. In contrast, steady state of crystal size distribution may take up to 12 residence times.

Thus, we chose to investigate the behavior of AgBr precipitations in a continuous suspension crystallizer prior to the establishment of crystallization steady state.[10] Dynamic mass balance and attenuation models for crystal number were used to describe the transient behavior prior to the renucleation point. Renucleation occurred when the crystal growth rate exceeded the critical growth rate. We observed that the time for the beginning of renucleation was affected by the precipitation conditions and was independent of the initial content of the reaction vessel. The time to renucleation varied from about 2.5 to 6.5 residence times, depending on both the precipitation conditions and the initial seeding conditions. For unseeded experiments, the initial crystal number was equal to that of a reference batch precipitation. It was observed that after the onset of renucleation, at least four additional residence times were needed for the continuous precipitation system to achieve a steady state.

Introduction

The precipitation of silver halide crystals in a continuous suspension crystallizer has been previously investigated.[9,11,12] Continuous streams of reactants, silver nitrate, halide salt, and gelatin solutions are fed into a well-stirred precipitation vessel, while product is simultaneously removed to maintain a constant reaction volume. A basic precipitation system for continuous chemical processes and crystallizations, the continuous suspension crystallizer is also known as controlled continuous stirred tank reactor (CSTR), or mixed-suspension, mixed-product-removal (MSMPR) reactor. For silver halide precipitations, the precipitation vessel initially contains gelatin and halide salt, and may include silver halide seed crystals. After a transient time period, a steady state is reached, after which the size distribution and shape of the silver halide crystals removed from the precipitation vessel remain unchanged. The system in which seed crystals are continuously added to the reaction mixture for continuous growth will be discussed in a separate section (A New Crystal Nucleation Theory).[8]

Randolph and Larson suggested that at steady state the dynamics and stability of a continuous suspension crystallizer are greatly influenced by the kinetic relationship between the nucleation rate (B°) and the linear growth rate (G) (Equation [9.4]), where i is a kinetic order relating B° to G.[13]

$$B^0 \alpha G^i \tag{9.4}$$

These authors concluded theoretically that for high-yield systems (e.g., silver halides), the crystallization process would be inherently stable if the kinetic order i was

1 τ   2.5 τ   4 τ

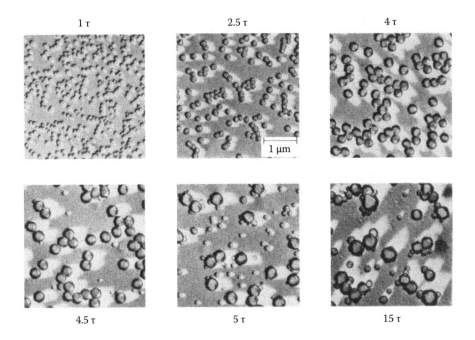

4.5 τ   5 τ   15 τ

FIGURE 9.4   Run 1, electron micrographs taken during the transient period of an unseeded continuous AgBr precipitation.

less than 21. Since i = 4 for the AgBr system, a steady state should be achieved without cycling of the crystal size distribution for AgBr precipitation in a continuous suspension crystallizer.[14] The stability of the AgBr system was confirmed experimentally by Wey and colleagues, who observed a time-invariant, steady-state crystal size distribution after about eight residence times into the precipitation.[9] However, before steady state was achieved, the crystallizer experienced non-steady-state transient behavior.

This non-steady-state transient behavior of an unseeded AgBr precipitation in a continuous suspension crystallizer is demonstrated in Figure 9.4, which shows typical electron micrographs of the crystallizer content taken during the first several residence times. The AgBr crystals are monodisperse at least until four residence times after the start of precipitation. At about 4.5 residence times, a second generation of considerably smaller crystals suddenly appears. As time goes on, the crystal population becomes more and more polydisperse and finally reaches a steady-state size distribution after about eight to nine residence times, and is shown for 15 residence times.

The initial transition phase was thus investigated for initially unseeded and seeded conditions.

MODEL FOR THE INITIAL TRANSIENT PHASE

To describe the initial transient behavior of the precipitation process in a continuous suspension crystallizer, the dynamic mass balance and a dynamic population

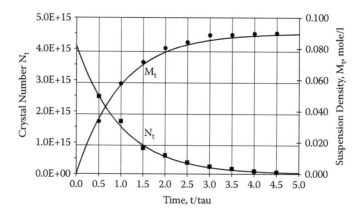

FIGURE 9.5   Run 1, AgBr, number, N_t, and suspension density, M_t, versus time.

balance were used. The dynamic population and mass-balance model of the steady state developed by Randolph and Larson[14] is basically a number-continuity equation in crystal-phase space and is generally difficult to solve.

However, for the initial portion of an unseeded AgBr precipitation, before the generation of the second crystal population, effective nucleation occurs only at the very beginning of precipitation, which is a phenomenon similar to that which occurs in a double-jet precipitation.[15] Thus, the crystals remain uniform in size distribution for several residence times, as illustrated in Figure 9.4. In this case, the dynamic population balance can be greatly simplified, and the total number of crystals (N) in suspension can be represented by a simple attenuation equation, (9.5).

$$N_t = N_0 e^{-t/\tau} \tag{9.5}$$

In Equation (9.5), N_0 is the total number of stable nuclei generated at the very beginning of precipitation. Otherwise, when monodisperse seed crystals are used in the initial precipitation vessel, N_0 represents the total number of seeds, t is the precipitation time, and τ is the average residence time in the precipitation vessel and is equal to the reaction volume (V) divided by the total volumetric flow rate (Q).

At the same time, the suspension density, M_t, increases with time for an unseeded precipitation as shown in Equation (9.6), where R is given in dmole/dt.

$$M_t = M_f(1.0 - e^{-t/\tau}) = \frac{V_m R \tau}{V_0}(1.0 - e^{-t/\tau}) \tag{9.6}$$

A dynamic mass balance based on total silver in the continuous AgBr precipitation system can be written as Equation (9.7).

$$V\frac{d}{dt}(M_t + C_\infty) = Q_i C_i - Q C_\infty - Q M_t \tag{9.7}$$

Here, M_t represents the AgBr crystal suspension density (mole/volume), C_∞ the bulk solute concentration, C_i the input reactant concentration, and Q_i the input reactant flow rate. The reactant mass addition rate, R (g/min), is equal to the product, Q_iC_i. If molar addition rate (i.e., mole/min) is chosen, R is equal to $V_mQ_iC_i$, where V_m is the molar volume of the crystals (g/mole).

For the high-yield AgBr system, it is reasonable to assume that accumulation of the dissolved reactant material in the bulk solution is negligible (i.e., $M_t \gg C_\infty$ and $C_i \gg C_\infty$). Hence, Equation (9.7) can be simplified to (9.8).

$$V\frac{dM_t}{dt} = R - QM_t \tag{9.8}$$

Equation (9.8) can be integrated to give

$$M_t = M_0e^{-t/\tau} + \frac{R\tau}{V}(1 - e^{-t/\tau}) \tag{9.9}$$

M_0 represents the suspension density of stable nuclei or seed crystals at the very beginning of the continuous precipitation.

For monodisperse crystal size distributions, M_t and M_0 can be given by Equations (9.10) and (9.11).

$$M_t = \frac{k_v\rho NL^3}{V} \tag{9.10}$$

$$M_0 = \frac{k_v\rho N_0L_0^3}{V} \tag{9.11}$$

Here, k_v is the crystal volume shape factor ($k_v = 1$ and 0.47 for cubes and octahedra, respectively), and ρ is the crystal density (moles/volume) and is equal to $1.0/V_m$; L is the crystal edge length, and L_0 is the edge length of the stable nuclei or seed crystals. The crystal edge length as a function of time can be determined from Equations (9.5), (9.9), (9.10), and (9.11).

$$L = \left[L_0^3 + \frac{R\tau}{k_v\rho N_0}(e^{-t/\tau} - 1) \right]^{1/3} \tag{9.12}$$

The total surface area (A) of the growing crystals becomes Equation (9.13).

$$A = k_aNL^2$$

$$A = k_aN_0e^{-t/\tau}\left[L_0^3 + \frac{R\tau}{k_v\rho N_0}(e^{-t/\tau} - 1) \right]^{2/3} \tag{9.13}$$

where k_a is the crystal area shape factor ($k_a = 6$ and 3.46 for cubes and octahedra, respectively). The linear growth rate (G) is inversely proportional to crystal surface area (A) and can be obtained from a simple mass balance equation, (9.14).

$$G = \frac{dL}{dt} = \frac{R}{3k_v\rho NL^2} = \frac{k_aR}{3k_v\rho A} \tag{9.14}$$

If the stable nuclei (or seed crystals) are very small ($L_0 \sim 0$), Equations (9.13) and (9.14) reduce to

$$A = K_A e^{-l/\tau}(e^{l/\tau} - 1)^{2/3} \tag{9.15}$$

$$G = K_g e^{l/\tau}(e^{l/\tau} - 1)^{-2/3} \tag{9.16}$$

where

$$K_A = k_a N_0^{1/3}\left(\frac{R\tau}{k_v\rho}\right)^{2/3} \tag{9.17}$$

$$K_g = \frac{k_a R}{3k_v\rho K} \tag{9.18}$$

Equations similar to (9.15)–(9.18) were also reported by Lieb and Osmers for crystallizations in a continuous suspension crystallizer with an initial pulse distribution.[16]

The transient behavior of the total crystal surface area and the linear growth rate for the case where $L_0 = 0$ is shown in Figure 9.6. The total surface area initially increases with time, reaches a maximum at 1.1 t/τ ($= \ln 3.0$), and subsequently decreases. The linear growth rate, which is inversely proportional to the total surface area, has a minimum at 1.1 t/τ.

This transient behavior can be understood and explained qualitatively by considering two key factors in such systems. On one hand, the surface area per crystal increases continually, because the stable nuclei (generated at the very beginning of an unseeded precipitation) or the seed crystals grow during precipitation. On the other hand, the total number of crystals decreases according to Equation (9.5), because of continuous removal (washout) of crystals through the output stream. Initially, the

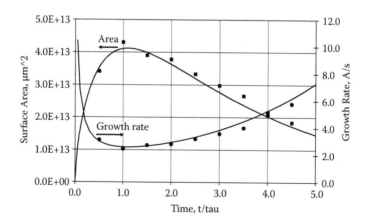

FIGURE 9.6   Run 1, total crystal surface area (A) and linear growth rate (G) versus time.

effect of crystal growth dominates so that the total crystal surface area increases and the linear growth rate decreases. This is qualitatively represented by the curves between 0 and 1.1 t/τ in Figure 9.6.

However, when the washout effect outweighs the surface area contribution from crystal growth, the total crystal surface area begins to decrease, and the linear growth rate starts to increase. This behavior is qualitatively indicated by the curves at $t/\tau > 1.1$ in Figure 9.6.

Finally, renucleation (formation of new nuclei) will occur at some point into the precipitation (at about 4.5 t/τ for the example shown in Figure 9.4) when the total crystal surface area becomes too small to accommodate the reactant material and the growth rate (or the supersaturation level) exceeds the critical value. After this point, effective nucleation continues to occur, and a steady state is eventually reached, as indicated by a time-invariant, polydisperse crystal size distribution (e.g., $t/\tau = 15$).

Although the transient behavior shown in Figure 9.6 is for the case where $L_0 = 0$, the same behavior is also expected for the cases where $L_0 \neq 0$. However, for the latter, Equations (9.13) and (9.14) should be used to calculate the surface area (A) and the growth rate (G). Furthermore, the maximum of area A and the minimum in G will occur at t/τ values different from 1.1.

EXPERIMENTAL

The equipment setup and operation of the continuous suspension crystallizer for AgBr precipitations have been described previously and in the section.[9] Two reactant solutions (AgNO$_3$ and NaBr) and an inert bone-gelatin solution were simultaneously introduced to an agitated reaction vessel. Both unseeded and seeded experiments were carried out to investigate the transient behavior (Table 9.1).

The resulting AgBr suspension was continually withdrawn to maintain a constant suspension volume. The precipitation conditions were kept constant for a given run.

TABLE 9.1
Experimental Conditions

Run	Seed Size, L_0 (μm)	Total Seed Number, N_0	Q_i (ml/min)	C_i (mole/l)	Input Gelatin[a] Flow Rate (ml/min)	V (ml)	τ (min)	Temp. (8C)	pBr
1	---	---	10	0.9	80	300	3.0	70	2.2
2	---	---	10	0.9	80	300	3.0	70	3.9
3[b]	---	---	10	0.9	80	300	3.0	70	3.9
4	---	---	5	4.0	50	1200	20.0	80	3.3
5	---	---	20	4.0	200	1200	5.0	80	3.3
6	0.07	3.34 × 1015	20	4.0	200	1200	5.0	80	3.3
7	0.06	1.56 × 1017	20	4.0	200	1200	5.0	80	3.3

[a] The input gelatin concentration was 2.4% for Runs 1–3 and 2.88% for Runs 4–7.
[b] AgBr(I) (5 mole % I).

By introducing seed crystals initially to the reaction vessel, one can independently control the N_0 and L_0 values and adjust the relationship between G and t/τ during the transient period for a given set of precipitation conditions. Such a technique is important for sorting out the important factor(s) responsible for controlling renucleation from experimental data.

Table 9.1 lists the experimental conditions. For the unseeded experiments (Runs 1–5), the reaction vessel was charged with a given volume of the input gelatin solution and was adjusted to the desired temperature and pBr prior to each precipitation. For the seeded experiments (Runs 6 and 7), cubic monodisperse AgBr seed crystals were added to the initial reaction vessel. Run 3 contained 5 mole% iodide in the halide reactant solution.

Samples were taken from the output stream at intervals of 0.5 t/τ during the transient period. The number of residence times at which renucleation occurred was determined for each experiment by observing the presence of small nuclei in electron micrographs of the samples. Both direct electron micrographs and a Joyce-Loebl disc centrifuge were used to measure the average crystal size and size distribution.[17] Crystal suspension density (M_t) was determined by analyzing the amount of silver halide present in the sample.

RESULTS AND DISCUSSION

The Transition Stage

The transient behavior shown in Figure 9.4 for Run 1 was examined in more detail. The average crystal size (L) and the crystal suspension density (M_t) for the samples taken up to the renucleation point were measured. Knowing L and M_t, one can calculate the total number of crystals (N_t) from Equation (9.10). In Figure 9.7, the number and size of the crystals during the start-up phase are plotted. While the number of crystals follows an exponential loss curve, the crystal size increases from near zero in a nonexponential correlation.

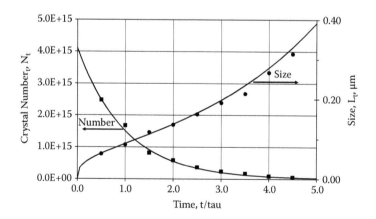

FIGURE 9.7   Run 1, crystal number and size versus time.

TABLE 9.2

Experimental Values of L and M_t and Calculated Values of N, A, and G as a Function of t/τ for Run 1

t/τ	L (μm)	M_t (mole/l)	N	A (μm²)	G (A/sec)
0.5	0.063	0.0335	2.48×10^{15}	3.41×10^{13}	3.12
1.0	0.086	0.0577	1.68×10^{15}	4.30×10^{13}	2.47
1.5	0.117	0.0714	8.25×10^{14}	3.90×10^{13}	2.73
2.0	0.136	0.0804	5.91×10^{14}	3.78×10^{13}	2.82
2.5	0.162	0.0843	3.67×10^{14}	3.33×10^{13}	3.20
3.0	0.191	0.0888	2.36×10^{14}	2.98×10^{13}	3.58
3.5	0.214	0.0890	1.68×10^{14}	2.66×10^{13}	4.01
4.0	0.267	0.0898	8.73×10^{13}	2.15×10^{13}	4.96
4.5	0.314	0.0900	5.38×10^{13}	1.84×10^{13}	5.79

The total crystal surface area (A) and the linear growth rate (G) were calculated from Equations (9.13) and (9.14), respectively. Table 9.2 lists the values of L, M_t, N, A, and G as a function of t/τ up to the renucleation point of Run 1. Plots of the total crystal surface area and the linear growth rate for this run are included in Figure 9.6. The transient behavior obtained from Run 1 is practically identical with that of the theoretical results shown by the solid curve in Figure 9.6. That is, the total crystal surface area (A) increased with time until about $1.1t/\tau$ and then decreased.

The linear growth rate (G), whose behavior is opposite to that of A, had a minimum at about 1.1 t/τ. Similar transient behavior was also found for the other unseeded experiments, Runs 2–5. Therefore, the theoretical considerations given earlier for the transient behavior were experimentally confirmed. The time at which renucleation occurred was about 4.5 t/τ. From these data, the critical growth rate for renucleation was calculated to be about 4.8 A/sec for Run 1.

As discussed earlier, during the initial portion of an unseeded continuous precipitation, effective nucleation occurs only at the very beginning, and the crystals are continuously being removed through the output stream. The total number of crystals (N_t) in suspension should follow the simple attenuation equation represented by Equation (9.5). Thus, a plot of ln N versus t/τ should yield a straight line with a slope of -1. From the intercept of this plot, one can also determine the total number of stable nuclei (N_0) generated from the initial nucleation step. Figure 9.8 shows such a plot for the experimental data obtained from Run 1.

The data correlate very well with a straight line. The slope of this line is −0.94 (+/−0.07) with the confidence range at the 98% level, and thus statistically indistinguishable from the theoretical value of −1.0. The N_0 value determined from the intercept of this plot is 3.9 (+/−0.8) $\times 10^{15}$.

Initial Nucleation versus Batch Nucleation

We also carried out a double-jet precipitation under the same conditions, i.e., molar addition rate of reactants, reaction volume, temperature, and pBr used in Run 1. The total number of stable nuclei obtained from this double-jet precipitation was

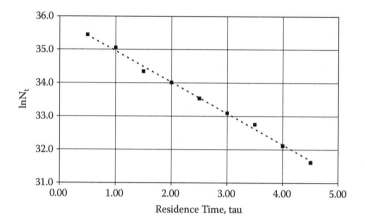

FIGURE 9.8   Run 1, AgBr, ln N_t versus time.

determined to be 3.5×10^{15}, which, within the limits of error, agrees with the N_0 value of 3.9×10^{15} obtained from Run 1.

The experimental values of t/τ and L and the calculated values of N (from Equation (9.10) and G (from Equations (9.13) and (9.14) at the renucleation point are listed in Table 9.3 for all experimental runs.

For the unseeded precipitations (Runs 1–5), renucleation occurred at about 3.5 to 5.5 t/τ for a variety of experimental conditions, including changes in reaction volume, reactant addition rate, temperature, pBr, halide type, and residence time, τ.

For the seeded precipitations (Runs 6 and 7), which had the same precipitation conditions, the time to reach the renucleation point varied from 2.5 to 6.5 t/τ, depending on the initial seeding conditions. For the conditions used in this study, the time to reach renucleation was affected more by the initial seeding conditions than by the precipitation conditions.

TABLE 9.3

Experimental Values of t/τ and L, and Calculated Values of N, G*, and S* at Renucleation

Run	t/τ	L (μm)	N	G* (A/sec)	S*	S_{ss}^*
1	4.5	0.31	5.38×10^{13}	5.8	---	---
2	3.5	0.25	4.94×10^{13}	4.7	---	---
3	5.5	0.27	4.33×10^{13}	4.6	---	---
4	4.0	0.64	4.14×10^{13}	1.9	1.29	1.42
5	4.0	0.37	1.97×10^{14}	4.3	1.41	1.94
6	2.5	0.34	2.66×10^{14}	4.2	1.37	1.94
7	6.5	0.36	2.43×10^{14}	4.1	1.38	1.94

[a] S_{ss} represents the supersaturation ratio at steady state reported in reference 3.

Maximum Growth Rate and Critical Supersaturation

The behavior of the growth rate at renucleation (G*) is of particular interest. For the unseeded precipitations (Runs 1–5), G* varied from about 1.9 to 5.8 A/sec, depending on the precipitation conditions. For example, G* increased from about 4.7 to 5.8 A/sec when pBr was decreased from 3.9 to 2.2 (Run 2 versus Run 1). On the other hand, the presence of iodide had no significant effect on G* (Run 2 versus Run 3). Finally, G* decreased from about 4.3 to 1.9 A/sec when τ was increased from 5 to 20 min (Run 5 versus Run 4). A further discussion of the above effects of pBr and iodide on G* was not possible at the time of the publication of the transition model, owing to a lack of quantitative information in the literature and a viable theory for the steady state. The latter was derived much later, relating average crystal size to the control variables, residence time, solubility, and temperature, and will be discussed in the following section.

The effect of τ on G* can be explained qualitatively. Wey and Strong reported that small AgBr crystals grew faster than large ones for a given supersaturation.[18] Since the crystal size at the renucleation point is larger for Run 4 ($\tau = 20$ min) than for Run 5 ($\tau = 5$ min), one would expect a smaller G* value for Run 4 than for Run 5. This consideration is qualitatively consistent with the experimental results and the quantitative steady-state model.

The effects of initial seeding conditions on the critical growth rate can be studied from Runs 6 and 7. Both runs had practically the same G* value despite their differences in the initial seeding conditions and in the t/τ value to reach the renucleation point. More interestingly, the G* values obtained from Runs 6 and 7 were essentially the same as that obtained from an unseeded precipitation (Run 5) with the same precipitation conditions.

The above results for G* show that renucleation occurs when the growth rate exceeds a critical value. This critical value can be influenced by the precipitation conditions (i.e., temperature, pBr, and τ) and, for a given set of precipitation conditions, is independent of the initial content (i.e., seeding conditions including the presence or absence of seed crystals) of the reaction vessel. Note: This is predicted by the BNG model and is experimentally confirmed there.

Although the critical growth rate (G*) can be conveniently used to explain and correlate the renucleation points for different experimental conditions, a more fundamental approach to the understanding of the renucleation phenomenon is to examine the critical supersaturation levels at the renucleation point. The critical supersaturation ratio (S*) for renucleation can be related to G* by a size-dependent growth rate expression for AgBr crystal growth.[21]

$$S^* = 1 + \frac{(1+\varepsilon L)G^*}{K_i C_e} \qquad (9.19)$$

At 70°C and pBr 3.3, the growth rate parameter, ε, and the kinetic integration constant, K_i, were determined to be about 16 μm^{-1} and 3.85×10^7 (A/sec)/(mole/1), respectively.[21] To a first approximation, these ε and K_i values may also be used for the conditions of 80°C and pBr 3.3. The equilibrium concentration (C_e) of AgBr crystals at 80°C and pBr 3.3 was calculated to be 1.9×10^{-6} mole/l. The S* values determined from Equation (9.18) for Runs 4–7 are listed in Table 9.3. Similar S* values for Runs 1–3 could not be determined, owing to a lack of knowledge of the growth rate parameters under these conditions. Runs 6 and 7, which had the same precipitation conditions but different initial seeding conditions, had essentially the same S* value (S* 1.38). This is close to the value of 1.41 obtained from the unseeded precipitation (Run 5) with the same precipitation conditions. On the other hand, a change in precipitation conditions influenced the S* value (Run 4 versus Run 5). For example, S* decreased from 1.41 to 1.29 as τ was increased from 5 to 20 min. Thus, the critical supersaturation ratio for renucleation appears to be influenced by the precipitation conditions but is independent of the initial (seeded or unseeded) content of the reaction vessel.

Wey et al. reported the calculated supersaturation ratio values at steady state (S_{ss}) for experiments carried out under the same precipitation conditions as Runs 4 and 5.[9] As shown in Table 9.3, the S* values are smaller than the corresponding S_{ss} values. That is, for a given continuous AgBr precipitation, the supersaturation level at steady state (where growth and nucleation rates are invariant with time and the two processes are occurring simultaneously) is higher than that at the renucleation point, as expected.

After the Renucleation Point

The model developed and the experimental results analyzed here mainly address the transient behavior up to the renucleation point. However, it is desirable to determine the number of additional residence times required for the continuous AgBr precipitation system to achieve a steady state after the renucleation point. Unfortunately, the theoretical treatment for this time period requires the solution of the complicated dynamic population balance equation,[8] owing to the simultaneous occurrence of nucleation and growth and the resulting polydisperse crystal size distribution.

Therefore, we experimentally analyzed the crystal size distribution of the samples taken after the renucleation point to determine the residence times after which the crystal size distribution remained unchanged. At $6\ t/\tau$, the distribution still had two modes, which represent crystals generated from the initial nucleation step and the renucleation step. At $8\ t/\tau$, the distribution had only one mode, but it is still significantly different from those of 9 and $15\ t/\tau$. The distributions of 9 and $15\ t/\tau$ are indistinguishable within the reproducibility of the measurement technique. The results clearly indicate that the steady state of Run 1 was approached between 8 and $9\ t/\tau$. Since renucleation occurred at about $4.5\ t/\tau$ for Run 1, it took at least four more residence times for the system to reach steady state after the onset of renucleation. We also examined the crystal size distributions for the other experimental runs and obtained similar results. That is, at least four residence times are needed after the renucleation point before the continuous AgBr precipitation system reaches a steady state.

Note: After the renucleation point, the number of renucleated crystals is determined by the maximum growth rate, the number of growing crystals, and the kinetic of crystal removal. It is anticipated that the renucleation theory, derived in Chapter 8, may be able to model the transition from the start of renucleation to the beginning of the steady state.

Summary

A theoretical and experimental study was carried out to examine the transient behavior of AgBr precipitation in a continuous suspension crystallizer. The crystal size distribution was monodisperse until renucleation occurred at several residence times into the precipitation. After the renucleation point, the AgBr crystal population became more and more polydisperse and finally reached a steady-state size distribution. The transient behavior prior to the renucleation point was successfully described by an attenuation model for crystal number and a dynamic mass balance.

For an unseeded AgBr precipitation, the total crystal surface area increased with time until about 1.1 t/τ and then decreased. The linear growth rate behavior was opposite to that of the total surface area and had a minimum at about 1.1 t/τ. The total number of crystals (N_t) followed the attenuation function, which yielded a straight line with a slope of about -1.0 for a plot of ln N versus t/τ. The total number of stable nuclei generated from the initial nucleation step of the continuous precipitation was experimentally determined to be equal to that obtained from a corresponding double-jet precipitation.

Renucleation occurred when the growth rate (or the supersaturation level) exceeded a critical value, which was affected by the precipitation conditions (i.e., temperature, pBr, and τ) and was independent of the initial content, i.e., the presence or absence of seed crystals of the reaction vessel. The time to reach the renucleation point was influenced by initial seeding conditions and to a lesser extent by precipitation conditions. The supersaturation ratio at the renucleation point was estimated to be smaller than that at steady state.

Renucleation was modeled in Chapter 8, and that model correlates renucleation with the reaction control parameters.

By analyzing the crystal size distribution of the samples taken after the renucleation point, we determined that at least four additional residence times are needed for the continuous precipitation system to achieve a steady state.

A NEW CRYSTAL NUCLEATION THEORY FOR SIZE CONTROL IN CONTINUOUS PRECIPITATIONS

SUMMARY

A new theory of crystallization in the continuous stirred tank reactor (CSTR, or mixed-suspension, mixed-product-removal, MSMPR) system was developed, which is based on the dynamic balance between growth and nucleation. The model is based on nonseeded systems with homogeneous nucleation and diffusion-controlled growth based on the nucleation model previously derived for such systems in controlled double-jet batch precipitations.[15]

The model predicts the correlation between the average crystal size, residence time, solubility, and temperature of the reaction system. It allows the determination of factors that are experimentally difficult to determine, such as the ratio of average to critical crystal size, L/L_c, the supersaturation ratio, S^*, the maximum growth rate, G_m, the ratio of nucleation to growth, R_n/R_i, and the size of the nascent nuclei, L_n.

While the present model was developed for homogeneous nucleation under diffusion-limited growth conditions and unseeded systems, it may be modified to model seeded systems, systems containing ripening agents or growth restrainers, and systems in which growth and nucleation are kinetically controlled.

INTRODUCTION

It is the object of this section to extend the previously reported model of crystal nucleation for precipitation of highly insoluble compounds in controlled double-jet precipitation batch processes[1,15,20,21] to precipitations in continuous suspension crystallizers.

The proposed new theory of crystallization in continuous mixed-suspension, mixed-product-removal (MSMPR) crystallizers, here referred to as continuous stirred tank reactor, CSTR, differs from the previous theory for continuous precipitation by Randolph and Larson (R-L)[14,22] in that it correlates the average crystal size of the crystal population with the reaction control parameters residence time, temperature, and crystal solubility. It is anticipated that the effect of ripeners and restrainers can be incorporated into this model.

Several aspects distinguish this model from the R-L model:

1. This model is based on a dynamic balance between crystal nucleation and crystal growth at steady state.
2. A mathematical treatment of the model is based on mass and nucleation balance. It shares certain concepts of mass and nucleation balance with the Randolph-Larson theory.
3. In addition, it introduces the concepts and equations of the crystal nucleation theory previously proposed and experimentally supported by the author and his coworkers.
4. This leads to new mathematical equations that have no arbitrary adjustable parameters. These new equations are significantly different from those of the R-L model.
5. Furthermore, the new theory and the equations make distinctly new predictions. Some of these were experimentally supported in the present work.
6. The new theory is supported by experimental results. As in any new endeavor, only few predictions could be tested. The theory and the equations give ample suggestions for further experimental and theoretical research.

THEORY

Preconditions

The Randolph-Larson model was reviewed in the first section (Continuous Crystallization). The second section will concentrate on the derivation of the new BNG-based model.

In continuous mixed-suspension, mixed-product-removal crystallizers, MSMPR, reactants, solvents, and other addenda are continuously added while the product is continuously removed. In the following, this precipitation scheme will be referred to as a continuously stirred tank reactor (CSTR) system.

For the present derivation, only the reaction-controlling reactant that leads to the crystal population will be considered. For instance, silver halides are generally precipitated with excess halide in the reactor, which is used to control the solubility. Thus, the soluble silver salt, in this case $AgNO_3$, is the reaction-controlling reactant. The solubility of the resultant silver halides (chloride, bromide, iodide, and mixtures) in the reactor is low so that their concentration in the reactor will be neglected for the mass balance.

Other necessary addenda for the precipitation, such as halide salt solutions, water, and gelatin, are included in the calculation of the residence time, τ, and suspension density, M_t, and are important to control the silver halide solubility. The present model will be based on several premises. It is apparent that modifications of the present model can be obtained by changing some of these premises.

1. The reactants stoichiometrically form the crystal population, and the solubility of the resultant product (in the present experiment, silver halide) is not significant with regard to the mass balance. Note: It is straightforward to expand the model to include the effect of significant solubility of the reaction product.
2. Homogeneous nucleation is assumed. For nonhomogeneous and other nucleation processes, the proper nucleation model must be substituted for the presently used nucleation model.
3. The input reactant stream at steady state is consumed in a constant ratio for crystal nucleation and crystal growth.
4. Crystal nucleation in the CSTR system follows the same mechanism as in double-jet precipitation.
5. Crystal growth is given by the maximum growth rate of the crystal population. In the present derivation, a single maximum growth rate, G_m, is assumed, which can be derived from the experimental results. However, an analytical equation for G_m as a function of crystal size and reaction conditions may be substituted if it is known. This reduces the number of unknowns in the final equation to the ratio of average to critical crystal size, L/L_c.
6. For the present derivation, it is assumed that the nucleation is by a diffusion-controlled mechanism.[25]

The following approach was successfully used previously to quantitatively describe renucleation in batch precipitations.[23]

MODEL DERIVATION

The basic reaction scheme for the following derivation is shown in Figure 9.9.

A reactant stream R_0 (addition rate of reactants, e.g., silver nitrate and alkali halide, mol/sec) is added to the reaction vessel at a steady rate. A fraction of R_0, R_n, will be used to nucleate new crystals to replace crystals leaving in the product. Part of the remainder of R_0, R_i, is used to sustain the maximum growth rate of the crystal population (addition rate fraction used for growth = crystal size increase). For

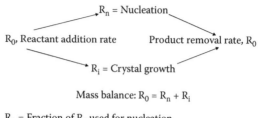

Mass balance: $R_0 = R_n + R_i$

R_n = Fraction of R_0 used for nucleation
R_i = Fraction of R_0 used for size increase (Growth)

FIGURE 9.9   BNG Model for continuous crystallization.

precipitations where the solubility of the reaction product is significant, the solubility and unreacted material (R_u) need to be added as a rate term on the right side of Equation (9.20).

The simplifying assumption that $R_u \ll R_n + R_i$ is well met by the precipitation of highly insoluble materials such as the silver halides that perform at practically 100% conversion. The mass balance requires that

$$R_0 = R_n + R_i + R_u \tag{9.20}$$

The addition rate R_0 is given by the concentration (C_r, mol/l) and flow rate (F_r, l/s) of the reactants.

$$R_0 = \sum_i C_{r,i} F_{r,i} \tag{9.21}$$

The product removal rate at steady state is by definition equal to the reactant addition rate. In the following, R_n and R_i will be derived and finally inserted into Equation (9.20) to provide the new model for the crystal population at steady state.

CRYSTAL NUCLEATION

The number of crystals nucleated, Z_n, must be equal to the number of crystals leaving in the reaction stream, and is given by the mass balance, Equation (9.22).

$$Z_n = \frac{R_0 V_m}{k_v L^3} \tag{9.22}$$

Here, V_m is the molar volume (cm³/mole of crystal mass) of the reaction product to convert molar addition rate into volume cm³/mol. The constant, k_v, is the volume factor that converts from the characteristic average crystal size, L, to crystal volume. L is used instead of r (used for batch precipitations) to indicate that it is the maximum crystal size for continuous processes.

In homogeneous crystal nucleation under diffusion-controlled growth conditions, Equation (9.23) was derived for double-jet precipitations.[1,7,15]

$$Z_n = \frac{R_n R_g T}{2 k_s \gamma D V_m C_s \Psi} \tag{9.23}$$

$$\text{Where} \quad \Psi = \frac{L}{L_c} - 1.0 \tag{9.24}$$

R_g is the gas constant, T the temperature (K), k_s the crystal surface factor, γ the surface energy, D the diffusion coefficient of the reaction-controlling reactant, and C_s the sum of the solubility with regard to the reaction-controlling reactant. L_c is the critical crystal size at which a crystal has equal probability to grow or to dissolve by Ostwald ripening.

In previous papers,[1,7,15] Equation (9.23) was quoted for the specific case of spherical crystal morphology ($k_s = 4\pi$). For batch processes, Z_n equals the total number of stable crystals formed during nucleation. In the present continuous precipitation, Z_n is equal to the nucleation rate (dZ/dt) at steady state. This extrapolation is in agreement with the underlying derivation of Equation (9.23).[1,15]

For the remainder of the derivation of the equations, the intermediate variable K is introduced (Equation [9.25]).

$$K = \frac{R_g T}{2 k_s \gamma D V_m C_s \Psi} \tag{9.25}$$

This leads to a simplified equation for Z_n:

$$Z_n = K R_n \tag{9.26}$$

Equation (9.26) is solved for R_n:

$$R_n = \frac{Z_n}{K} \tag{9.27}$$

Substitution of Z_n from Equation (9.22) into the equation for R_n (Equation [9.27]) leads to Equation (9.28):

$$R_n = \frac{R_0 V_m}{K k_v L^3} \tag{9.28}$$

Thus, an analytical equation for R_n has been found, which will be entered into Equation (9.20) (see Crystal Growth and Nucleation, below). It remains now to develop an analytical equation for R_i, which will be derived from the maximum growth rate of the crystal population.

CRYSTAL GROWTH

The maximum growth rate, G_m, is given by the mass balance between the maximum growth of the system and the fraction of the reactant addition rate consumed for this growth, R_i. The maximum growth rate is a function of crystal size, and thus the individual size classes will grow at different absolute maximum growth rates. The maximum growth rate of the whole crystal population is then given by an average maximum growth rate, G_m, which is defined to be related to the average crystal size, L.

If the crystal size is difficult to determine because of a complicated crystal structure, for instance dendritic crystals, the specific surface area, S_m (i.e., surface area/mole of crystals), may be used to derive the maximum growth rate. This was not necessary under the present conditions. The use of S_m was discussed by Leubner[23,30] and may be transferred to the present model as desired.

Equation (9.29) results from solving this mass balance for G_m.

$$G_m = \frac{V_m R_i}{3.0\, k_v L^2\, Z_t} \tag{9.29}$$

Here, Z_t is the total number of crystals present in the reaction vessel during steady state. Z_t can be calculated from the average crystal size and the suspension density, which, in turn, is a function of reactant addition rate and residence time.

$$Z_t = \frac{R_0 V_m \tau}{k_v L^3} \tag{9.30}$$

Inserting Equation (9.30) into Equation (9.29) and solving for R_i leads to the desired analytical equation for R_i:

$$R_i = \frac{3 G_m R_0 \tau}{L} \tag{9.31}$$

At this point, both R_n and R_i have been expressed by analytical expressions that contain only fully defined parameters and variables. It is now possible to continue to the formulation of the new theory.

CRYSTAL GROWTH AND NUCLEATION IN THE CONTINUOUS CRYSTALLIZER

The foundation has now been laid to combine nucleation and growth at steady state. For this purpose, the expressions of R_n (Equation [9.28]) and R_i (Equation [9.31]) are inserted into Equation (9.20). After simplifying, Equation (9.32) is obtained:

$$\frac{V_m}{Kk_v L^3} + \frac{3 G_m \tau}{L} = 1.0 \tag{9.32}$$

Reinserting K into this equation and solving for zero leads to:

$$k_v L^3 R_g T - 2 k_s \gamma DV_m^2 C_s \Psi - 3 k_v G_m R_g TL^2 \tau = 0 \tag{9.33}$$

We have now derived an equation that relates the average crystal size at steady state for the CSTR crystallization to the reaction control variables τ, residence time, T, temperature, and C_s, solubility. Other factors in Equation (9.33) that are experimentally available are R_g, the gas constant; k_s, the crystal surface constant, which converts the characteristic crystal size to crystal surface; k_v, the crystal volume constant, which converts the characteristic crystal size to crystal volume; V_m, the molar volume (cm^3/mole) of the crystals; and D, the diffusion constant of the reaction-controlling ion or growth entity.

The only unknown factors are G_m, the maximum crystal growth rate, and Ψ, from which the ratio of average to critical crystal size, L/L_c, is obtained (Equation [9.24]). The unknown factors G_m and Ψ are experimentally obtained from Equation (9.33).

Further, the model predicts that the crystal size and number are independent of the reaction volume.

PRACTICAL ANALYSES

Solving for the average size, L, results in a very complicated solution as a function of residence time, which is of little immediate use. Thus, another method is used to determine the average size as a function of reaction control variables.

For instance, solving for the residence time, τ, is relatively straightforward and is given by Equation (9.34). The method for variation of solubility, C_s, and temperature, T, follows the same procedure.

$$\tau = \frac{L}{3G_m} - \frac{2k_s \gamma D V_m^2 C_s \Psi}{3\, k_v G_m R_g T\, L^2} \tag{9.34}$$

For practical calculations and for conditions where one or the other constant is not known, Equation (9.34) can be simplified to Equation (9.35)

$$\tau = aL - \frac{b}{L^2} \tag{9.35}$$

where

$$a = \frac{1.0}{3G_m} \tag{9.36}$$

$$b = \frac{2k_s \gamma\, DV_m^2 C_s \Psi}{3k_v G_m R_g T} \tag{9.37}$$

$$\frac{b}{a} = \left(\frac{2k_s \gamma D V_m^2 C_s}{k_v R_g T} \right) \Psi \tag{9.38}$$

Equations (9.35)–(9.38) will be used to determine G_m and Ψ from the experimental values of residence time, τ, and the average crystal size, L, obtained at crystal population steady state. In addition, other parameters will be entered that are either known or can be experimentally determined.

If G_m is known from other experiments, or an analytical equation of G_m is available that relates it to the crystal size and reaction conditions, this may be substituted in Equation (9.34). This would reduce the number of unknowns in determining Ψ.

From Ψ and the average crystal size, L, the critical crystal size, L_c, can be calculated using Equation (9.24). L_c is related to the supersaturation, S*, ratio by Equation (9.39).

$$S^* = \frac{(C_{ss}-C_s)}{C_s} = 1.0 + \frac{2\gamma V_m}{R_g T L_c} \tag{9.39}$$

Here, C_{ss} is the actual concentration (supersaturation), and C_s is the equilibrium solubility as defined above.

SPECIAL LIMITING CONDITIONS FOR CONTINUOUS CRYSTALLIZATION

Two limiting cases of Equation (9.34) are of special interest. For this purpose, Equation (9.34) is rewritten

$$\tau = \frac{L}{3G_m}\left(1.0 - \frac{2k_s\gamma DV_m^2 C_s \Psi}{k_v L^3 R_g T}\right) \tag{9.40}$$

Large Average Crystal Size, L

If the average crystal size is very large as given by the definition in Equation (9.41), then the bracket of Equation (9.40) can be set equal to one and simplified to Equations (9.42) and (9.43).

$$L^3 \gg \frac{2k_s\gamma DV_m^2 C_s \Psi}{k_v R_g T} \tag{9.41}$$

$$\tau = \frac{L}{3G_m} \tag{9.42}$$

$$L = 3G_m\tau \tag{9.43}$$

Equation (9.43) indicates that for large sizes, L and τ, and L are linearly related and that G_m, the maximum growth rate, can be determined from the linear part of the correlation at large crystal sizes. G_m can then be resubstituted into Equations (9.36)–(9.38) to determine Ψ.

Short Residence Time, Plug-Flow Reactor, $\tau \rightarrow$ zero

Equations (9.34) and (9.35) indicate that when τ, the residence time, approaches zero the average crystal size will not become zero but will reach a limiting value. Plug-flow reactors are characterized by short residence times, $\tau \rightarrow 0$, in the nucleation zone, followed by crystal growth and ripening processes. The present derivation predicts the minimum crystal size for the condition where $\tau \to 0$:

$$L^3 = \frac{2k_s \gamma D V_m^2 C_s \Psi}{k_v R_g T} \tag{9.44}$$

$$V_g = k_v L^3 \tag{9.45}$$

Substituting into Equation (9.44), where V_g is the average crystal volume of the crystals, leads to Equation (9.46).

$$V_g = \left(\frac{2k_s \gamma D V_m^2 \Psi}{R_g} \right) \frac{C_s}{T} \tag{9.46}$$

Thus, for residence times approaching zero, the average crystal volume, Equation (9.46) predicts that V_g is only a function of solubility, C_s, and temperature, T. The factor Ψ may be a function of C_s and T.

CALCULATION OF REACTION PROCESSES

Nucleation versus Growth, R_n/R_i

In the model used for these derivations, the reactant addition stream R_0 is separated in the reactor into a nucleation stream, R_n, and a growth stream, R_i (Figure 9.9). It is now possible to derive the ratio of the reactant streams R_n/R_i by dividing Equation (9.28) by (9.31). After back-substituting for K (Equation [9.25]), Equation (9.47) is obtained:

$$\frac{R_n}{R_i} = \frac{2k_s \gamma D V_m^2 C_s \Psi}{3k_v G_m R_g T \tau L^2} \tag{9.47}$$

Except for τ in the divisor, this is equal to the second part on the right side of Equation (9.34). If we simplify to (9.48) it becomes evident that the ratio R_n/R_i decreases with increasing residence time, τ, and with the square of the average crystal size at steady state.

$$\frac{R_n}{R_i} \sim \frac{1}{\tau L^2} \tag{9.48}$$

It appears also intuitive that the nucleation should decrease relative to growth, as the residence time and the surface area of the crystal population (proportional to L^2) increase.

The value of the ratio R_n/R_i was calculated from the experimental data. From the ratio R_n/R_i and the mass balance (Equation [9.21]), R_n and R_i were calculated as a fraction of the total reactant addition rate, R_0.

Nascent Nuclei Size

For the present work and in agreement with common usage, the term *nascent nuclei*, L_n, is defined as the stable crystals that are newly formed during steady state and that continue to grow in the reactor. The size of these crystals is larger than that of the critical nuclei, L_c, which have equal probability to grow or dissolve in the reaction mixture.

With the model that has been developed up to this point, it is possible to calculate L_n for the different precipitations. For this purpose we define L_n by

$$L_n^3 = \frac{R_n V_m}{k_v Z_n} \tag{9.49}$$

Here, Z_n can be calculated from Equation (9.22), and R_n can be calculated from Equations (9.21) and (9.47). Back-substitution into Equation (9.49) leads to Equation (9.50):

$$L_n^3 = \left(\frac{R_n}{R_0}\right) L^3 \tag{9.50}$$

This equation was used to calculate L_n values for the different residence times.

It is now possible to back-substitute further from Equation (9.50). Equation (9.51) is obtained, which relates the nascent nuclei size to the precipitation conditions and the average crystals size at steady state:

$$L_n^3 = \frac{2k_s \gamma D V_m^2 C_s \Psi L^3}{\left(3k_v G_m R_g T \, \tau L^2 + 2\,k_s \gamma D V_m^2 C_s \Psi\right)} \tag{9.51}$$

The terms in this equation have been defined above. It is apparent that L_n is a complicated function of the reaction conditions and of the average crystals size, which in itself is a complicated function of the same reaction variables (Equation [9.33]).

CONCLUSIONS

A new theory of crystallization is proposed for the continuous stirred tank reactor (CSTR) or mixed-suspension, mixed-product-removal (MSMPR) system. The model is based on nonseeded systems with homogeneous nucleation, diffusion-controlled growth, and the nucleation model previously derived for such systems in controlled double-jet batch precipitations. It does not need any assumptions about size-dependent growth (McCabe law).

The model predicts the correlation between the average crystal size and the residence time, solubility, and temperature of the reaction system. It allows the determination of useful factors that are experimentally difficult to determine, such as the ratio of average to critical crystal size, L/L_c, the supersaturation ratio, S^*, the supersaturation, C_{ss}, the maximum growth rate, G_m, and the ratio of nucleation to growth, R_n/R_i.

The model predicts that the average crystal size is independent of reactant addition rate, suspension density, and reaction volume.

The model predicts that the average crystal size and residence time are linearly related when the average crystal size is significantly larger than a certain limiting size, which can be derived from the experimental correlations. At smaller crystal sizes, the average crystal size is larger than predicted by the linear part of the correlation. These predictions were supported by experimental results in the following section.

The model further predicts that when the residence time approaches zero, the average crystal size approaches a limiting value larger than zero. The condition where the residence time approaches zero is similar to that obtained during nucleation in plug-flow reactors and thus predicts a lower limit of average crystal size for the CSTR and plug-flow systems.

The present model may be expanded to include the effect of Ostwald ripening agents and growth restrainers using the formalism previously applied to precipitations in batch precipitations.[2,3]

The present model was developed for homogeneous nucleation under diffusion-limited growth conditions and unseeded systems. It may be modified to model systems in which growth and nucleation are kinetically controlled. Modeling of crystal growth in continuously seeded systems will be presented in the following section. Most importantly, the present work suggests many additional experiments for and new approaches to the evaluation of the results of continuous precipitations.

SIZE DEPENDENCE ON RESIDENCE TIME IN CONTINUOUS PRECIPITATIONS

SUMMARY

Continuous precipitations of silver chloride with varying residence times from 0.5 to 5.0 min were chosen to support predictions of the BNG theory of crystallizations in the continuous stirred tank reactor (CSTR) developed in the previous section. The experiments confirm the model prediction that the average crystal size is independent of reactant addition rate, suspension density, and reaction volume. However, the experiments indicate that the width of the crystal size distribution increases with suspension density. The model confirms that the average crystal size and residence time are linearly related for large average crystals. The model also predicts, and is confirmed experimentally, that when the residence time approaches zero (plug-flow condition), the average crystal size approaches a positive limiting value. The condition where the residence time approaches zero (0.205 µm) is equal to that predicted for nucleation in plug-flow reactors, and thus predicts a lower limit of average crystal size for these systems. The size of the nascent (newly formed) crystals at steady state, L_n, was determined to be 0.205 +/– 0.012 µm for the residence time range from 0.5 to 5.0 min. This is equal to the zero time condition. The ratio of the fraction of the input stream used for nucleation, R_n, to the fraction used for crystal growth, R_i (R_n/R_i), varied from 4.79 to 0.12 for residence time range from 0.5 min to 5.0 min. The ratio of average to critical crystal size, L/L_c, was determined to 5.73 × 10^3 (1.02–1.09), the supersaturation ratio, S^*, to 12.2 (0.54, $L = 0.5$ µm), the supersaturation to 8.2 × 10^{-8} (12.7 × 10^{-9}, $L = 0.5$ µm), and the maximum growth rate, G_m, to 4.68 A/s.

INTRODUCTION

It is the object of this section to test the predictions of the BNG model for size control in precipitations in continuous suspension crystallizers for the effect of residence time on the average crystal size for silver chloride precipitations. The equations used for the evaluation of the experimental results were presented in the previous section and will not be repeated in this section.

The use of continuous precipitation systems for the precipitation of silver halide dispersions has been investigated previously. Gutoff,[11,12] Wey and Terwilliger,[24] and Wey and colleagues[9,25] investigated the crystal size distribution using the formalisms derived by Randolph and Larson for the mixed-suspension, mixed-product-removal (MSMPR) system.[14,22] Wey, Leubner, and Terwilliger[10] examined the crystal size distribution of AgBr using the population balance technique and including both the size-independent McCabe's ΔL law and a size-dependent growth model. The crystal population distribution could not be satisfactorily modeled. Using the large grain population only, nucleation and maximum growth rates were determined using the Randolph-Larson model.

Wey and Jagannathan[27] studied the transient behavior of unseeded silver bromide precipitations and determined that the steady state of crystal population distribution was achieved only at about six to nine residence times (τ) after the start of the precipitations, much later than the steady state of suspension density (at about four residence times). Their results also showed that at steady state the crystal population is rather narrowly distributed around an average crystal size.

EXPERIMENTAL

The present experiments were done before the present theory was developed. If the theory had been known at the time of the experiments, a wider range of experiments would have been performed to determine the present results to a greater degree. Unfortunately, the author is no longer in a position to provide additional experiments. However, the present experiments support several important predictions of the theory and might be the starting point for more extended work in the future.

Silver chloride, AgCl, was precipitated in a single-stage continuous stirred tank reactor (CSTR) system. The residence time was varied from 0.5 to 5.0 min (Table 9.4).

TABLE 9.4

Effect of Residence Time on the Average Crystal Size of AgCl Precipitated in the CSTR System

Exp. no.	Residence Time, τ (min)	Edge Length, L (μm)	Size Distribution, d.r.	Crystal no. ($Z_t \times 10^{12}$)	Reactor Volume, V_0 (ml)	Flow Rate, F (ml/min)	Addition Rate, R_0 (mol/min)	Suspension Density, M_t (mol/l)
1	0.5	0.207	2.69	73	300	50	0.050	0.083
2	1.0	0.263	2.64	36	300	25	0.025	0.083
3	3.0	0.337	2.58	41	600	20	0.020	0.100
4	5.0	0.413	2.22	37	1000	20	0.020	0.100

Reaction conditions: 60°C; AgNO$_3$, NaCl 1.0 mol/l; 2.4% bone gelatin; τ, residence time (min); V_0, reactor volume (l); F, flow rate (ml/min, AgNO$_3$ and NaCl); R_0, molar addition rate (mol/min); M_t, suspension density (mol AgCl/l); L (μm), cubic edge length; d.r., decade ratio (measure of size distribution); Z_t, total crystal number in reactor.

TABLE 9.5
Effect of Suspension Density on Average Crystal Size and Number for Precipitations of AgCl in the CSTR System

Exp. no.	Suspension density, M_t (mol/l)	Edge length, L (μm)	Size distribution, d.r.	Crystal no. (Z_t x 10^{12})	Reactant concentration, C (mol/l)	Addition rate, R_0 (mol/min)
1	0.05	0.327	2.45	11	0.5	0.005
2	0.10	0.350	2.59	18	1.0	0.010
3	0.10	0.337	2.58	41	1.0	0.020
4	0.20	0.331	2.79	43	2.0	0.020
5	0.30	0.333	(3.51)[a]	63	3.0	0.030
6	0.40	0.344	3.25	81	4.0	0.040

Reaction conditions: 60°C; pAg 6.45; residence time, τ, 3.0 min; all experiments except no. 3: $AgNO_3$, NaCl 10 ml/min; gelatin (2.4%) 80 ml/min; V_0, 300 ml.

Experiment no. 3: $AgNO_3$, NaCl 20 ml/min; gel (2.4%) 160 ml/min; V_0, 600 ml.

C is the concentration of reactants ($AgNO_3$ and NaCl); R_0, molar addition rate (mol/min); M_t, suspension density (mol AgCl/l); L, average crystal size (cubic edge length, μm); d.r., decade ratio (measure of crystal size distribution); Z_t, total crystal number in reactor.

[a] Data not reliable.

For the residence time of 3.0 min, the suspension density was varied from 0.05 to 0.40 mol/l (Table 9.5).

The reactor volume, flow rates, and reactant concentrations are given in Tables 9.4 and 9.5. Bone gelatin was used as the peptizing agent. The temperature was held constant at 60°C. The free silver ion activity, {Ag^+}, was controlled in the reactor at pAg 6.45, where pAg = -log {Ag^+}. This corresponds to a solubility of 6.2×10^{-6} mol Ag^+/l.[26] This solubility consists of the sum of concentrations of free silver ion plus complexes of silver ions with halide ions, $AgCl_n^{1-n}$ (n = 1 – 4). The silver chloride precipitated in cubic morphology. A crystal growth restrainer was added to the output material to avoid Ostwald ripening and to preserve the crystal size distribution.

The crystal size distribution of the crystal suspensions was determined using the Joyce-Loebl disc centrifuge.[17,23] The original data were determined as equivalent circular diameter (ecd) and were converted to cubic edge length (cel), where cel = 0.86 × ecd. The crystal size distribution curves were determined in the ecd scale. The crystal size distribution was fitted to the sum of two Gaussian distributions, and thus the distribution cannot be described by a single standard deviation. Thus, the crystal size distribution is given by an empirical measure, the decade ratio (d.r.), which is defined by the ratio of the sizes at 90% to 10% of the experimentally determined crystal size population.

Electron micrographs (carbon replica) of AgCl crystals precipitated in the CSTR system (5 min residence time) are shown in Figure 9.10. The crystal size distributions

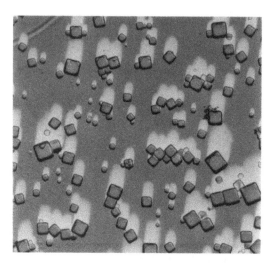

FIGURE 9.10   Electron micrographs of silver chloride crystals (carbon replica) obtained at steady state. Residence time 5.0 min, 60°C, pAg 6.45, 2.4% gelatin, suspension density 0.1 mole/l.

for 0.5 to five minute residence time are shown in Figure 9.11 and the experimental conditions and results in Table 9.4. The suspension density effect on crystal size, number, and crystal size distribution was studied for the 3.0 min residence time (Table 9.5). The constants used for the calculations are listed in Table 9.6.

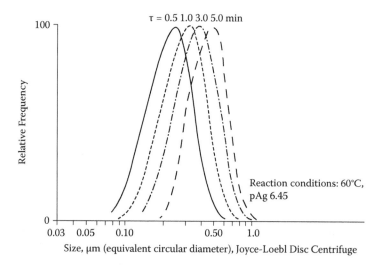

FIGURE 9.11   Silver chloride crystal size distribution for residence times varying from 0.5 to 5.0 min. (60°C, pAg 6.45, 2.4% gelatin).

TABLE 9.6
Constants Used in the Calculations

Constant	Value	Comment
k_v	1.0	cubic
k_s	6.0	cubic
γ	52.2	ergs/cm^2
D	1.60×10^{-5}	cm^2/s
V_m	25.9	cm^3/mol AgCl
C_s	6.2×10^{-9}	mol/cm^3
R_g	8.3×10^7	erg/deg - mol
T	333 K/60°C	

RESULTS AND DISCUSSION

The new model makes a number of predictions that were tested with the experimental results:

1. The average crystal size is independent of addition rate and suspension density.
2. The size-dependence of the average crystal size on residence time can be modeled using the equations given.
3. The average crystal size has a final value at zero residence time.
4. The maximum growth rate, the critical crystal size, the supersaturation ratio and the supersaturation, and the ratio of nucleation to growth of the system can be determined at steady state.

CRYSTAL SIZE, ADDITION RATE, AND SUSPENSION DENSITY

The model predicts that the average crystal size is independent of the reactant addition rate and by implication of the suspension density and reaction volume. This is supported by the results in Table 9.5 and Figure 9.12, which show that the average crystal size did not significantly vary over a range of addition rates from 0.005 to 0.04 mol/min and a range of suspension density from 0.05 to 0.40 mol/l. At the same time, the total crystal number in the reactor, Z_t, increased proportionally to the molar addition rate.

Experiments 2 and 3 have the same suspension density but vary by a factor of 2 in molar addition rate; however, the average crystal size and the decade ratio are not significantly different. In these experiments, the doubling of the addition rate leads to a doubling of the total crystal number.

Similarly, Experiments 3 and 4 have the same molar addition rate but vary in the suspension density by a factor of 2, while producing the same average crystal size and total crystal number.

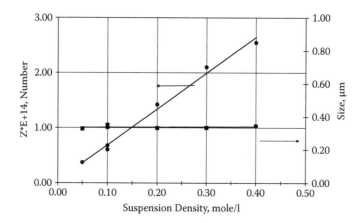

FIGURE 9.12   Crystal size and number as a function of suspension density.

The invariance of crystal size with suspension density is confirmed by the statistics in Equation (9.52). The linear correlation of crystal number versus suspension density is similarly supported by Equation (9.53). In summary, these results confirm the prediction of the theory that the crystal number is independent of suspension density and by implication independent of the population density.

$$\text{Size} = 0.337 \pm 0.009 \ \mu m \tag{9.52}$$

$$Z = (6.66 \pm 0.09)10^{14} * \text{Suspension Density(mole/l)} \tag{9.53}$$

All error limits equal one standard deviation; $R^2 = 0.9956$

Table 9.5 shows that the width of the crystal size distribution, as measured by decade ratio (d.r.), increased with increasing suspension density. This is indicated by Experiments 2 and 3 where for Experiment 3 the molar addition rate (and reactor volume V_0) was doubled while the suspension density was held constant. The value of d.r. is the same, indicating that the suspension density, not the molar addition rate, affects the width of the crystal size distribution.

Experiments 3 and 4 have the same addition rate, but Experiment 4 has half the reactor volume and flow rate, so that it has twice the suspension density of Experiment 3. The experiment with the higher suspension density (Experiment 4) has the wider crystal size distribution. This reinforces the direct relationship between suspension density and width of crystal size distribution. The unusually wide size distribution of Experiment 5 is probably due to some undetermined experimental deviation.

The crystal size distribution is governed by two different reactions: For crystals that are larger than the stable crystal size, growth is dominated by the maximum growth rate. This part of the crystal population can probably be described by the Randolph-Larson model, which is based on the maximum growth rate of the crystals.

The crystals that are smaller than the stable crystal size also grow at maximum growth rate, but also disappear at some rate by Ostwald ripening. Ostwald ripening is the process by which larger crystals increase in size ("ripen") at the expense of the dissolution of smaller ones. The critical crystal size at which a crystal has equal probability to grow or dissolve by Ostwald ripening can be determined by the present model and experiments from the factor Ψ.

In addition, it was determined that in controlled double-jet precipitations, the maximum growth rate nonlinearly decreases with increasing crystal size.

It was shown that the maximum growth rate increases under crowded conditions when the crystal population density is very high and where the diffusion layers of the crystals overlap.[23,27,28] This effect may depend on the crystal size and the state of supersaturation in the reactor and thus may contribute to the increase in crystal size distribution with increasing suspension density.

Since two different reaction mechanisms, growth and the Ostwald ripening effect, are differently effective for the different fractions of the crystal size populations, it can be anticipated that the shape of the resultant crystal population might not be symmetrical.

The result that the average crystal size is independent of molar addition rate and suspension density allows adding the result from Table 9.5 to those of Table 9.4 for the correlation of crystal size with residence time.

CRYSTAL SIZE AND RESIDENCE TIME

The results from Tables 9.4 and 9.5 were combined and plotted in Figure 9.13. A linear least squares evaluation using L and $1.0/L^2$ based on the predicted size/τ correlation results in Equation (9.54), where τ is in minutes and L in μm.

$$\tau = (11.88 \pm 0.77)L - \frac{(0.1026 \pm 0.0214)}{L^2}$$

$$R^2 = 0.984$$

(9.54)

The error limits (one std. deviation) show that the model and experimental results are in agreement. Further, the results show that at low residence times, the crystal size does not decrease to zero.

In Figure 9.13, the solid line is given by Equation (9.54). The linear correlation (dashed line) represents the linear L/τ correlation for large crystal sizes.

MAXIMUM GROWTH RATE, G_M

From Equation (9.54), the maximum growth rate was calculated to $G_m = 28.1 \times 10^{-3}$ μm/min, or 4.68 A/s. This is in good agreement with the results by Strong and Wey, who determined the maximum growth rate of AgCl in controlled double-jet precipitations to between 4.25 and 1.20 A/s for grain sizes between 0.209 and 0.700 μm.[26] This suggests that the growth rate, G_m, may be obtained independently from continuous precipitations to be used for the calculations in the present theory. In the double-jet precipitations G_m can be determined as a function of crystal size,

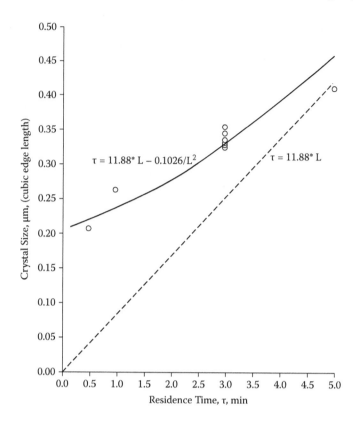

FIGURE 9.13   Crystal size of silver chloride crystals (μm cubic edge length) as a function of residence time (min), 60°C, pAg 6.45, 2.4% gelatin.

since the crystal size distribution is very narrow. In the present precipitations, G_m is the average of the maximum growth of a relatively wide crystal size distribution.

Minimum Crystal Size (τ->0)

This value of 0.205 μm is essentially the same as determined for $\tau = 0.5$ min, 0.207 μm.

Determination of Ψ and L/L$_c$, Supersaturation Ratio, S*, and Supersaturation, C$_{ss}$

The parameters in Table 9.6 were used to calculate Ψ (5.73×10^3) and L/L$_c$ (~5.73×10^3), the ratio of average to critical crystal size. The accuracy of the calculated results is affected by the constants in Table 9.6, especially uncertainties of the values of surface energy, γ, diffusion coefficient, D, and solubility, C_s.

In Table 9.7, the critical crystal size, L_c, the supersaturation ratio, S*, and the supersaturation, C_{ss}, during steady state are listed as a function of residence time, τ.

TABLE 9.7
Critical Crystal Size, Supersaturation, and - Ratio as a Function of Residence Time

Residence Time (min)	Cubic Edge Length (μm)	Critical Crystal Size, L_c (μm $\times$ 10^{-5})	Supersaturation Ratio, S*	Supersaturation, C_{ss} (mol/cm^3 $\times$ 10^{-8})
0.5	0.207	3.61	28.1	18.0
1.0	0.263	4.59	22.3	14.5
3.0	0.337	5.88	17.6	11.6
5.0	0.413	7.21	14.6	9.65

$\Psi = 5.73 \times 10^3$; for experimental conditions see Table 9.1.

The data show that the critical crystal size increases with residence time while the supersaturation and supersaturation ratio decrease.

If one adds 1.0 mol/l silver nitrate to a solution where the solubility is 6.2×10^{-6} mol/l, as in the present experiments, a theoretical supersaturation ratio of 1.60×10^5 is calculated at the silver nitrate entry point.[1] In stop-flow experiments, Tanaka and Iwasaki estimated that the size of the primary nuclei formed in AgCl precipitations was about $(AgCl)_8$.[29] Thus, the difference between the theoretical (1.60×10^5) and actual supersaturation ratio (14.6 – 28.1) in this system indicates that Ostwald ripening involving metastable nanoclusters may play a significant role in the nucleation/growth mechanism.

The supersaturation data, S*, for the present CSTR system are substantially higher than those reported for controlled batch double-jet precipitations under the same conditions (Table 9.8) as reported by Leubner[30] and Strong and Wey,[26] who reported values of S* of approximately 1.02 and 1.09, and values for L/L_c of 0.54 and 4.2. For comparison of the systems, an average crystal size of 0.5 μm was used.

TABLE 9.8
Calculated Experimental Constants and Variables for CSTR and Batch Precipitations of AgCl

Variable	CSTR	Batch	Comment	Variable	Ref. no.
Max. growth rate	4.68	1.20–4.25	A/s	G_m	b
Aver./critical size ratio	5.73×10^3	0.54	---	L/L_c	a
Supersaturation ratio	12.2	1.02–1.09	for L = 0.5 μm	S*	a, b
Supersaturation	8.2×10^{-8}	12.7×10^{-9}	mol/cm^3	C_{ss}	

$\Psi = 5.73 \times 10^3$; References for batch precipitations: (a) I. H. Leubner,[30] (b) R. W. Strong and J. S. Wey.[26]

The higher values of L/L_c and S^* for the CSTR versus the batch system indicate that during steady state, the balance between maximum growth and renucleation stabilizes much smaller critical crystal sizes than the batch double-jet precipitation. This is also supported by the wider crystal size distribution in the CSTR system.

However, the absence of very small crystals in the product suggests that upon removal of the high supersaturation in the product stream the small initial crystals rapidly dissolve by Ostwald ripening to produce the observed crystal size distributions. This is in agreement with simulations of the effect of Ostwald ripening on the crystal size distribution in batch precipitations by Tavare.[31] Unfortunately, this author did not provide experimental evidence for his simulations. The effect of Ostwald ripening and growth restraining agents on the crystal nucleation in batch precipitations was modeled and experimentally supported by the author.[2,3]

NUCLEATION VERSUS GROWTH, R_N/R_I

The model was used to calculate the ratios, R_n/R_i, and the values of R_n and R_i (as a percent of R_0) as a function of residence time, τ, and average crystal size, L. The results are listed in Table 9.9. The data show that at short residence times, nucleation (high R_n) is dominant, while at long residence times, growth (high R_i) dominates the reaction.

NASCENT NUCLEI SIZE

The sizes of the nascent nuclei, L_n, were calculated and are included in Table 9.9. The sizes of the nascent nuclei are independent of residence time with an average size of $0.205 +/- 0.012 \mu m$. This size range is the same as determined for the limiting crystal size for the plug-flow reactor, $0.205 \mu m$.

As a consequence of the relatively stable size of the nascent nuclei, their size relative to the steady-state average crystal sizes decreases with increasing residence time. Thus, at the 0.5 min residence time, the size of the nascent nuclei is about

---

TABLE 9.9

AgCl Crystal Nucleation and Crystal Growth Reactant Rates, and Nascent Nuclei Sizes

Residence Time (min)	Cubic Edge Length, L (μm)	R_n/R_i	R_n (% of R_0)	R_i (% of R_0)	L_n (μm)	% L_n (% of L)
0.5	0.207	4.79	82.7	17.3	0.194	93.9
1.0	0.263	1.48	59.7	40.3	0.221	84.2
3.0	0.337	0.30	23.1	76.9	0.207	61.4
5.0	0.413	0.12	10.7	89.3	0.196	47.5

R_n = fraction of reactant input rate consumed for crystal nucleation, R_i = fraction consumed for crystal growth at steady state. L_n = size of nascent nuclei, $0.205 +/- 0.012 \mu m$, %L_n = % of nascent nuclei sizes versus average size. Reaction conditions: AgCl, 60°C, pAg 6.45, 2.4% gelatin.

93.9% of the final crystal size, while for 5.0 min residence time, it is only about 47.5% of the final crystal size.

CONCLUSIONS

The BNG theory for continuous crystallization was tested for continuous precipitations of silver chloride, where the residence time was varied from 0.5 to 5.0 minutes. Reaction addition rates, reaction volume, and suspension density were varied. Solubility, temperature, and gelatin concentrations were held constant.

The BNG model predicts that the average crystal size is independent of reactant addition rate, suspension density, and reaction volume. This was supported with experiments at 3 minute residence time where the molar addition rate was varied from 0.005 to 0.040 mol/min, the suspension density from 0.05 to 0.40 mol/l, and the reaction volume from 300 to 1,000 ml. Unexpectedly, the width of the crystal size distribution (given by the decade ratio) increased with suspension density from 2.45 to 3.25, but was independent of reactant addition rate and reaction volume.

The model predicted that when the residence time approaches zero, the average crystal size approaches a limiting value larger than zero. For the present AgCl precipitations, the limiting value for zero minute residence time was calculated to about 0.205 μm. Between the residence times 0.5 and 5.0 minutes, the average crystal size varied from 0.207 to 0.413 μm. The condition where the residence time approaches zero is similar to that obtained during nucleation in plug-flow reactors and thus predicts a lower limit of average crystal size for the CSTR and plug-flow systems of about 0.205 μm.

The model also allowed calculating the fractions of the input reactant stream, R_0, that are used for nucleation, R_n, and for growth, R_i. The ratio, R_n/R_i, decreased with increasing residence time from 4.79 to 0.12. This indicates that with increasing residence time, more of the reactant input stream is consumed for growth than for nucleation.

The average crystal size of the nascent (newly formed) crystals, L_n, was independent of all reaction conditions that were varied, with an average value of 0.205 +/ −0.012 μm. This size is the same as extrapolated for the plug-flow condition ($\tau \to 0$) of 0.205μm.

It should be noted that most of the information quantitatively obtained from the BNG model and experiments has not been available from previous models and theories.

In the following section, the BNG model will be tested for the effect of crystal solubility on the crystal size in continuous silver chloride precipitations.

SIZE DEPENDENCE ON SOLUBILITY IN CONTINUOUS PRECIPITATIONS

SUMMARY

Nano crystals of silver chloride in the range 340–520 nm were prepared by varying the crystal solubility in a controlled continuous CSTR (MSMPR) crystallizer. The BNG model was used to correlate the average size with solubility. The solubility of silver chloride was varied from 0.81 to 8.3E-06 mol/l (60°C, 3.0 min residence time) with a correlation coefficient of 0.9996 between model and crystal size.

The experiment and model calculated the average maximum crystal growth rate (0.85 nm/s), the ratio of critical to average crystal size to 0.39, the nascent crystal size to 130–300 nm; critical crystal size to 130–200 nm, supersaturation ratio to 1.0049–1.0075, and the reactant split ratio between nucleation (R_n) and growth (R_i) to R_n/R_i from 0.06 to 0.26.

It was shown in the previous section that the average crystal size is independent of reactant addition rate, suspension density, and reaction volume.

INTRODUCTION

In crystallization processes and for the control of crystal size, growth, and design, the control of crystal nucleation is a crucial first step in research, product development, and manufacturing. For this purpose, the balanced nucleation and growth model (BNG) was developed.[1] In the previous sections, we have shown how the BNG model provides a fundamental model to control crystal nucleation and growth in continuous crystallizations. The strength of the BNG model is based on engineering control variables and the properties of the precipitate.[4,7,14,22] In this section, experimental results are presented for the control of crystal size by varying the crystal solubility to support the model.

THE BNG MODEL FOR THE CSTR SYSTEM

The present alternate model for size control in continuous crystallizations is based on the balanced nucleation and growth (BNG) model. This model quantitatively correlates the maximum crystal size at steady state with the precipitation temperature, product solubility, and reactant addition rate (Equation [9.55]). This model was supported by experiments where the residence time, τ, was varied.[8]

$$k_v R_g L^3 T - 2k_s \gamma D V_m^2 \Psi C_s - 3k_v R_g G_m L^2 T \tau = 0 \qquad (9.55)$$

G_m and Ψ may be functions of T, C_s, and τ. Ψ is defined in Equation (9.56), where L is the average crystal size. L_c is the critical crystal size, which in the crystallization system has equal probability to grow or dissolve.

$$\Psi = \frac{L}{L_c} - 1.0 \qquad (9.56)$$

The complexity of Equation (9.55) does not allow determining both G_m and Ψ for a single reaction condition. The determination of G_m and Ψ for a given variable/size combination may be achieved if the aim condition is bracketed by the symmetrical condition. Average values of G_m and Ψ can be calculated for a set of adjacent determinations, and trends of these parameters may be obtained. Because of the limited number of experiments available in the present and previous work, only average values of G_m and Ψ were obtained.

SIZE CONTROL BY CRYSTAL SOLUBILITY

The Solubility Model

Equation (9.55) was solved to Equation (9.57) to correlate solubility, C_s, with crystal size, L. The same method was previously applied to model the correlation between residence time and crystal size.[8]

$$C_s = \left(\frac{k_v R_g T}{2k_s \gamma DV_m^2 \Psi} \right)\left(L^3 - 3G_m \tau L^2\right) \qquad (9.57)$$

$$\text{Limits} \quad L \geq L_m, C_s \geq C_m$$

Equation (9.57) implies that the solubility can achieve a value of zero. However, Figure 9.14 shows that for AgCl, the solubility goes through a minimum, C_m, and does not equal zero at any condition.

In addition, the minimum size, L_m, is limited by the minimum solubility, and only the minimum and larger sizes will be correctly modeled. Nevertheless, Equation (9.57) allows the calculation of a finite solubility for the condition L = 0.

To give a meaningful correlation for the case presented for AgCl (Equation [9.57]), the solubility for the condition L = 0, C_0, must be added (Equation [9.58]). This addition is forced by the correlation of solubility on the second and third orders of size and the solubility property of AgCl, which does not include $C_s = 0$.

$$C_s = C_0 + \left(\frac{k_v R_g T}{2k_s \gamma DV_m^2 \Psi} \right)\left(L^3 - 3G_m \tau L^2\right) \qquad (9.58)$$

Equation (9.58) be can simplified by introducing an intermediate constant, K (Equations [9.59] and [9.60]). K is related to Ψ, which may be dependent on solubility. Thus, K may vary with solubility in relationship to Ψ. As discussed above, in

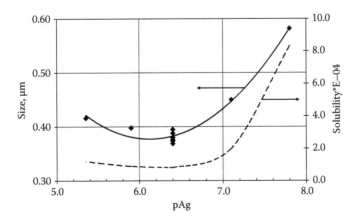

FIGURE 9.14 AgCl Solubility and crystal size versus pAg (-log[Ag⁺]).

the present experimental case, maximum growth rate, G_m, and Ψ values are average values over the solubility range, and thus K is a constant for these considerations.

$$C_s = C_0 + KL^3 - 3G_m \tau KL^2 \tag{9.59}$$

where

$$K = \frac{k_v R_g T}{2k_s \gamma DV_m^2 \Psi} \tag{9.60}$$

To solve Equation (9.59), a multivariate model was used to correlate crystal size to solubility.

$$C_s = a_0 + a_1 L^3 + a_2 L^2 \tag{9.61}$$

Here, a_0 represents the extrapolated solubility for $L = 0$, C_0, and a_1 is equal to K.

The fit of data to Equation (9.59) provides K, from which Ψ is obtained by solving Equation (9.60) for Ψ (Equation [9.62]).

$$\Psi = \frac{k_v R_g T}{2k_s \gamma DV_m^2 K} \tag{9.62}$$

The factors a_1 and a_2 were used to determine the average maximum growth rate, G_m (Equation [9.63]).

$$G_m = \frac{a_2}{3\tau a_1} \tag{9.63}$$

Two limiting conditions can be derived from Equation (9.57). The first is extrapolation to zero solubility (Equation [9.64]). This approximation is only allowed if C_s experimentally approaches zero. Under these conditions (which are not met by the solubility properties of AgCl), this extrapolation indicates that the size would have a limiting value for zero solubility.

For the AgCl system, the lowest solubility is greater than zero, and thus for the minimum solubility condition a size larger than L_0 is predicted.

$$L \rightarrow L_0 = 3G_m \tau \tag{9.64}$$

Another extrapolation is obtained for the condition given in Equation (9.65).

$$\text{For } L \gg 3G_m \tau \tag{9.65}$$

$$L^3 = \frac{2k_s \gamma DV_m^2 \Psi}{k_v R_g T} *C_s = \frac{C_s}{K} \tag{9.66}$$

This extrapolation (Equation [9.66]) predicts that at large crystal sizes (as defined), L increases linearly with solubility.

TABLE 9.10

Constants for AgCl Solubility Calculation

Component	$K_{m,n}$ (25°C)	ΔH (kcal/mol)
$[Ag^+]$	10^{-pAg}	
$[AgCl]$	$10^{-3.3}$	−2.7
$\left[AgCl_2^-\right]$	$10^{-5.25}$	−3.9
$\left[AgCl_3^{2-}\right]$	$10^{-5.7}$	−5.8
$\left[AgCl_4^{3-}\right]$	$10^{-5.4}$	−13.9

$$K_{sp} = [Ag^+][Cl^-] \qquad K_{m,n} = \frac{[Ag^+]^m[Cl^-]^n}{[Ag_mCl_n]^{m-n}}; m = 1$$

$$\log K_{sp} = -(3206 \pm 42)/T(K) + (1.17 \pm 0.14)$$

$$pAg = \frac{(599 - mV) - (T(C) - 25)*0.129}{(273 + T(C))*0.198}$$

$$T(K) = 273.15 + degC; T(C) = °C, mV = EMF$$

SOLUBILITY

The solubility of crystals is given by the sum of the concentrations of all soluble species in solution that contain the reaction-determining ion, and that participate in the growth of the crystals.

In the case of silver chloride, the precipitation is generally performed with chloride excess, so that the silver ion is the reaction-determining ion. Under these reaction conditions, the concentration of growth-determining silver ion, the solubility of the AgCl crystals, C_s, is given by the concentrations of free silver ion, of silver chloride molecules in solution, and of silver chloride, $AgCl_m^{1-m}$, complexes in solution as shown in Equation (9.67).

$$C_s = [Ag^+] + [AgCl] + \left[AgCl_2^-\right] + \left[AgCl_3^{2-}\right] + \left[Ag_4^{3-}\right] \tag{9.67}$$

The necessary constants to calculate the concentrations of these species are available from the literature and are compiled in Table 9.10.[32,33]

EXPERIMENTAL

The experimental design for a standard precipitation is shown in Table 9.11. Variations from these conditions are listed in Table 9.12. The steady state for crystal size and size distribution is only established after eight to 12 residence times, and the samples for size determination were pulled after 12 residence times.[10] Longer reaction times,

TABLE 9.11
Experiment: CSTR Precipitation Procedure for AgCl

300 ml Reaction volume (CSTR crystallizer)
2.4% Gelatin in distilled water
60°C
EMF = 170 mV, silver versus silver/silver chloride (4.0 N KCl) reference electrode
3.0 min Residence time
10 ml/min Reactant flow rate (AgNO3, NaCl)
1.0 mol/L Reactant concentration
80 ml/min Gelatin flow rate
0.10 mol/L AgCl Suspension density
Samples were collected after 12 residence times
Crystal size analysis by Joyce-Loebl disc centrifuge
The solubility was adjusted with sodium chloride to aim pAg

up to 50 residence times, did not show significant changes from these results. The experimental results are included in Table 9.12.

The solubility was varied by adjusting the pAg (-log [Ag$^+$]) in the reactor with sodium chloride. The pAg, which represents the activity of free silver ions in solution, was measured via the electromotive force (EMF) versus a reference electrode as indicated in Table 9.12. The size and size distribution was determined using a Joyce-Loebl disk-centrifuge system.[17] The reported sizes, which are given in equivalent circular diameter, were converted to cubic edge length by multiplying with the geometric factor 0.86.

TABLE 9.12
Experimental Variations, Crystal Size, and Number

pAg	Reaction Volume (mL)	Reactant Concentr. (Mol/L)	Reactant Flow Rate (mL/min)	Suspension Density (M/L)	Size (cel, μm)	Crystal no. (Z × E +14)
7.8	300	1.0	10	0.100	0.516	0.19
7.1	300	1.0	10	0.100	0.399	0.41
6.4	300	0.5	10	0.050	0.327	0.37
6.4	600	1.0	20	0.100	0.337	0.68
6.4	300	1.0	10	0.100	0.350	0.60
6.4	300	2.0	10	0.200	0.331	1.42
6.4	300	3.0	10	0.300	0.333	2.10
6.4	300	4.0	10	0.400	0.344	2.55
5.9	300	1.0	10	0.100	0.353	0.59
5.4	300	1.0	10	0.100	0.369	0.52

Results

Suspension Density

It was shown in the previous section discussing the effect of residence time, τ, that crystal size and number are independent of suspension density. The appropriate experiments are included in the present data.

Size-Solubility Correlation

Equation (9.57) predicts that in the CSTR crystallizer, the crystal size depends on the solubility, and the crystal number is varied by varying the suspension density. In contrast, for batch precipitations, the crystal number depends on the solubility, and the crystal size is a function of the amount of precipitated mass.[30]

In Figure 9.14, crystal size and solubility are plotted versus pAg. This plot shows that crystal size and solubility are related. For instance, both solubility and crystal size go through a minimum at about pAg 6.4, and are higher at lower and higher pAg.

To derive the correlation between size and solubility from the BNG model, the experimental data in Table 9.12 were analyzed using multivariate analysis according to Equation (9.61). According to Equation (9.59), a_0 is equal to the condition where $L = 0$. The results of the modeling are shown in Equation (9.68). The standard deviations and the correlation coefficient indicate excellent correlation of the data with the model.

$$C_s = (44.4 \pm 5.53) * 10^{-6} + (2.607 \pm 0.155) * 10^9 L^3 - (1.199 \pm 0.0996) * 10^5 L^2$$

$$\text{L (cm) cubic edge length (cel), } C_s \text{ (mol/l); } R^2 = 0.9996$$

(9.68)

The plot of the experimental data and the modeling fit, size versus solubility, are shown in Figure 9.15. The plot is in agreement with the model, which predicts that at low concentrations the size does not approach zero (Equation [9.64]). Further, as

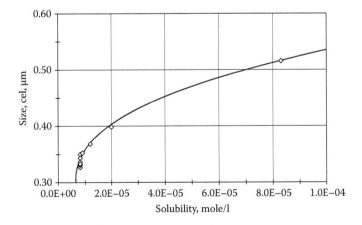

FIGURE 9.15 Model: Crystal size and model fit versus solubility.

TABLE 9.13

Constants for Model Calculations

Constant	Value	Comment
k_v	1.0	Cubic, volume constant
k_s	6.0	Cubic, surface constant
γ	52.5	erg/cm^2
D	1.60E-05	cm^2/s
V_m	25.9	cm^3/mol, AgCl
R_g	8.31E07	erg/deg × mol
	1.987	cal/deg × mol
T	333 K/60°C	Temperature

predicted by the model, the correlation approaches linearity with solubility at larger sizes (Equation [9.65]).

The necessary constants to calculate the derived parameters are listed in Table 9.13.

The coefficient of L^3, a_1, is equal to K, and this allowed calculating an average value for Ψ of 1.57 (Equation [9.62]). From Ψ, an average of L_c/L to 0.389 was obtained. The coefficients of L^3, a_1, and L^2, a_2, were used to calculate the average maximum growth rate, G_m, to 8.52 A/s (Equation [9.63]). The results of the model calculation, the average maximum growth rate (G_m), average stability constant (Ψ), S* factor, and reactant split factor (R_n/R_i), are collected in Table 9.14.

Since the average crystal size, L, is available from the experiment, the critical crystal size is calculated for each reaction condition according to Equation (9.69).

$$L^* = 0.389*L \tag{9.69}$$

From L_c, the critical crystal size, S*, the supersaturation ratio, was calculated for each reaction condition.[62] The S*- and R_n/R_i factors were used to calculate S* and R_n/R_i for each experiment.

TABLE 9.14

Calculated Model Parameters

Parameter	Value	Comment
G_m	8.52 (A/s)	Maximum growth rate
Ψ	1.57	
L_c/L	0.389	Critical/average size
R_n/R_i Factor	8.34E-6 × (C_s/L^2)	Ratio of nucleation to growth
S* Factor	1.0 + (9.83E-08)/L_c	Supersaturation ratio

TABLE 9.15

Derived Parameters: Critical, L_c, and Nascent, L_n, Crystal Size; Supersaturation Ratio, S*; and Reactant Split Ratios, R_n/R_i, R_n/R_0, R_i/R_0

pAg	Solubility (mol/l)	Size (cel, μm)	L_c (μm)	L_n (μm)	S*	R_n/R_i	R_n/R_0	R_i/R_0
7.8	8.30E-05	0.516	0.201	0.305	1.0049	0.260	0.207	0.793
7.1	2.00E-05	0.399	0.155	0.182	1.0063	0.105	0.095	0.905
6.4	8.10E-06	0.337	0.131	0.129	1.0075	0.060	0.056	0.944
6.4	8.10E-06	0.327	0.127	0.128	1.0077	0.063	0.059	0.941
6.4	8.10E-06	0.350	0.136	0.131	1.0072	0.055	0.052	0.948
6.4	8.10E-06	0.331	0.129	0.128	1.0076	0.062	0.058	0.942
6.4	8.10E-06	0.333	0.130	0.128	1.0076	0.061	0.057	0.943
6.4	8.10E-06	0.344	0.134	0.130	1.0073	0.057	0.054	0.946
5.9	9.00E-06	0.353	0.137	0.136	1.0072	0.060	0.057	0.943
5.4	1.20E-05	0.369	0.143	0.151	1.0069	0.074	0.069	0.931
pAg 6.4	Average	0.337	0.131	0.129	1.0075	0.060	0.056	0.944
	Std. dev.	0.009	0.003	0.001	0.0002	0.003	0.003	0.003

The experimental and the modeling results are compiled in Table 9.15 for all experiments. This includes the calculated results as a function of solubility of critical, L_c, and nascent, L_n, crystal size, the supersaturation ratio, S*, and the reactant split ratios, R_n/R_i, R_n/R_0, and R_i/R_0.

Experimental, Critical, and Nascent Crystal Sizes, L, L_c, and L_n

The experimental, critical, and nascent crystal sizes, L, L_c, and L_n, are plotted versus solubility in Figure 9.16. All crystal size types increase with increasing solubility. The critical and nascent crystal sizes are significantly smaller than the average

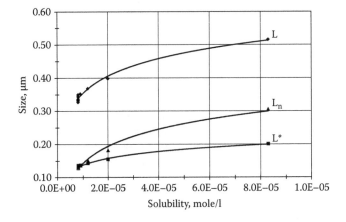

FIGURE 9.16 Size, L, critical size, L_c, and nascent size, L_n, versus solubility.

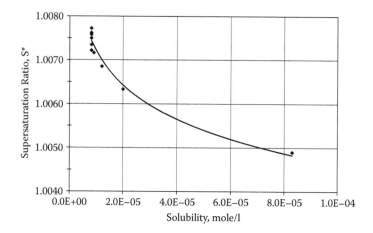

FIGURE 9.17   Supersaturation ratio, S*, versus solubility.

crystal size. At the lowest solubility, L_c and L_n are virtually equal. Above the minimum solubility, the nascent crystal size, L_n, is larger than the critical crystal size, L_c, and their separation increases with increasing solubility. The critical crystal size, L_c, has equal probability to grow or dissolve under the reaction conditions. Thus, the nascent crystal size, L_n, has a higher than equal probability to grow. The modeling results are thus in agreement with the expectations.

Supersaturation Ratio, S*

The supersaturation ratio, S*, is plotted versus solubility in Figure 9.17. The value of S* decreases with increasing solubility.

The supersaturation ratio for the experiments spans a range of 1.0049 to 1.0077. Lower supersaturation ratios were associated with higher solubility. The range indicates that the supersaturation is 0.49 to 0.77 percent higher than the saturation solubility.

Reactant Split Ratios, R_n/R_i, R_n/R_0, and R_i/R_0

The reactant split ratios, R_n/R_i, R_n/R_0, and R_i/R_0, are shown as a function of solubility in Figure 9.18 and in Table 9.15. The plot shows that the reactant split increases with increasing solubility. This indicates that at higher solubility, more reactant goes into nucleation and less into crystal growth. This may be related to the observation that the nascent crystal size, L_n, increases with solubility, and thus more of the incoming reactant stream is needed for nucleation (Figure 9.18).

Conclusions

The extension of the BNG model to the CSTR crystallizer resulted in an equation that relates crystal size and reaction conditions and is free of arbitrary parameters.

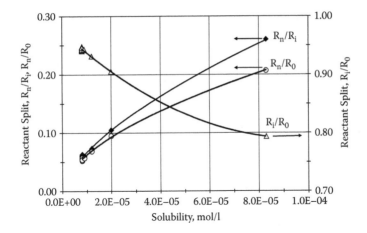

FIGURE 9.18   Reactant split ratios, R_n/R_i, R_n/R_0, and R_i/R_0 versus solubility.

In the present study, the correlation between crystal size and solubility was determined for the continuous precipitation of silver chloride. The values of two parameters, which are initially not known, Ψ, a factor related to the supersaturation ratio, and G_m, the maximum growth rate, are obtained from the combination of model and experiments. The value of Ψ allows calculating the supersaturation during steady state, and thus Ψ has a defined physical meaning. Individual values of G_m and Ψ for each solubility condition may be obtained by selectively designing solubility conditions. In the context of the present experiments, such data were not available, and average values of Ψ and G_m were obtained.

For the silver chloride system, the average maximum growth rate, G_m, was determined to 8.52 A/s. The average of the parameter Ψ is 1.57. It is directly related to the ratio of average to critical crystal size, L/L_c, from which the critical crystal size, L_c, was obtained (range 0.137–0.201 μm). The critical crystal size, L_c, is related to the supersaturation at steady state, which was calculated to vary between 1.0049 and 1.0075. L_c increases with solubility.

The model allows calculating the size of the nascent crystals, L_n, which are the initial stable crystals during the nucleation process. The size of L_n was found to increase with solubility from 0.129 to 0.305 μm. As anticipated, L_n is larger than L_c.

The reactant split ratio of incoming reactant used for nucleation and growth could be determined using the present model. The ratio of R_n/R_i increased from 0.060 to 0.260 with increasing solubility. From this ratio, the fraction of R_n and R_i versus R_0 (the incoming reaction stream) could be calculated (R_n/R_0 and R_i/R_0).

The predicted third-order correlation between solubility and crystal size was confirmed (correlation coefficient R = 0.9996). Since the AgCl solubility goes through a minimum and does not include zero solubility, an intercept was included, which represents the solubility where the crystal size is extrapolated to zero (L = 0). The zero-size condition cannot be obtained for the precipitation of AgCl.

The present experiments confirm the predictions of the BNG-based model for continuous crystallization for the correlation between crystal size and crystal solubility. These results add to the previous confirmation of the predictions for the size dependence on residence time, τ. Together, these confirmations of the BNG-based continuous crystallization model make it a useful model to correlate crystal size to variations of precipitation conditions.

REFERENCES

1. Leubner, I. H. 1987. Crystal formation (nucleation) under kinetically and diffusion controlled growth conditions. *J Phys Chem* 91:6069.
2. Leubner, I. H. 1987. Crystal formation (nucleation) in the presence of growth restrainers. *J Crystal Growth* 84:496.
3. Leubner, I. H. 1987. Crystal formation (nucleation) in the presence of Ostwald ripening agents. *J Imaging Sci* 29:31.
4. Bransom, H. S., W. J. Dunning, and B. Millard. 1949. *Disc Faraday Soc* 5:83.
5. Randolph, A. D., and M. A. Larson. 1962. *AICHE J* 8:639.
6. Randolph, A. D., and M. A. Larson. 1991. *Theory of particulate processes*. 2nd ed. San Diego: Academic Press.
7. Leubner, I. H. 1998. A new crystal nucleation theory for continuous precipitation of silver halides. *J Imaging Sci Technol* 42:355.
8. Leubner, I. H. 2001. Seeded precipitations in the continuous stirred tank reactor crystallizer. *J Disp Sci Technol* 22:373.
9. Wey, J. S., J. P. Terwilliger, and A. D. Gingello. 1980. AICHE Symposium Series, vol. 76, no. 193, 34.
10. Wey, J. S., I. H. Leubner, and J. P. Terwilliger. 1983. Transient behavior of silver bromide precipitation in a continuous suspension crystallizer. *Photogr Sci Eng* 27:35–39.
11. Gutoff, E. B. 1970. Nucleation and crystal growth rates during the precipitation of silver halide photographic emulsions. *Photogr Sci Eng* 14:248.
12. Gutoff, E. B. 1971. Nucleation and crystal growth rates during the precipitation of silver halide photographic emulsions. Pt. 2. Ammoniacal emulsions. *Photogr Sci Eng* 15:189.
13. Randolph, A. D., and M. A. Larson. 1971. In *Theory of particulate processes*. New York: Academic Press.
14. Randolph, A. D., and M. A. Larsen. 1962. Transient and steady-state size distributions in continuous mixed suspension crystallizers. *AIChE J* 8:639.
15. Leubner, I. H., R. Jagannathan, and J. S. Wey. 1980. Formation of silver bromide crystals in double-jet precipitations. *Photogr Sci Eng* 24:103.
16. Lieb, E. B., and H. R. Osmers. 1974. Paper presented at the Joint Meeting of GVC and AIChE, Munich, Germany; Lieb, E. B., and H. R. Osmers. 1973. The effect of diffusion on crystal size distribution in a continuous crystallizer. *Chem Eng Commun* 1973:17–20.
17. King, T. W., S. M. Shor, and D. A. Pitt. 1981. Technique for the characterization of size distribution during Ostwald ripening of silver halide crystals. *Photogr Sci Eng* 25:70.
18. Wey, J. S., and R. W. Strong. 1977. *Photogr Sci Eng* 21:14.
19. Wey, J. S. 1981. *Preparation and properties of solid state materials*. Ed. W. R. Wilcox. Vol. 6. New York: Marcel Dekker.
20. Leubner, I. H. 2001. The balanced nucleation and growth model for controlled crystallization. *J Disp Sci Technol* 22:125–38.
21. Leubner, I. H. 2002. The balanced nucleation and growth model for controlled crystal size distribution. *J Disp Sci Technol* 23:577–590.

22. Randolph, A. D., and M. A. Larson. 1988. *Theory of particulate processes, analysis and techniques of continuous crystallization.* 2nd ed. Academic Press, second edition, 1991.

23. Leubner, I. H. 1993. Crystal growth and renucleation, theory and experiments. *J Imaging Sci Technol* 37:357.

24. Wey, J. S., and J. P. Terwilliger. 1974. On size-dependent crystal growth rates. *AIChE J* 20:1219.

25. Wey, J. S., J. P. Terwilliger, and A. D. Gingello. 1976. Precipitation of silver halide emulsions in a continuous reactor. *Res Discl* 14:987.

26. Strong, R. W., and J. S. Wey. 1979. The growth of AgCl crystals in gelatin solution. *Photogr Sci Eng* 23:344.

27. Wey, J. S., and R. Jagannathan. 1982. Determination of growth kinetics of polyhedral crystals. *AIChE J* 28:697.

28. Jagannathan, R. 1988. Growth rate studies on cubic and octahedral AgBr crystals. *J Imaging Sci* 32:100.

29. Tanaka, T., and M. Iwasaki. 1985. The multistage process of formation of ultrafine silver bromide particles as revealed by multichannel spectroscopy. *J Imaging Sci* 29:20.

30. Leubner, I. H. 1985. Formation of silver halide crystals in double-jet precipitations: AgCl. *J Imaging Sci* 29:219.

31. Tavare, N. S. 1987. Simulation of Ostwald ripening in a reactive batch crystallizer. *AIChE J* 33:152.

32. Chateau, H., J. Pouradier, and C. R. Berry. 1971. In *The theory of the photographic process.* Ed. T. H. James. 3rd ed. New York: Macmillan, p. 6; Pouradier, J., et al., 1954. *J Chem Phys* 51:375; Towns, M. B., et al. 1960. *J Phys Chem* 64:1861; Greeley, W., et al. 1960. *J Phys Chem* 64:652.

33. Pouradier, J., A. Pailliotet, and C. R. Berry. 1977. In *The theory of the photographic process.* Ed. T. H. James. 4th ed. New York: Macmillan.

10 Crystal Growth in the Continuous Crystallizer

SEEDED PRECIPITATIONS IN THE CONTINUOUS STIRRED TANK REACTOR CRYSTALLIZER

SUMMARY

Growth of crystals in a continuous stirred tank reactor (CSTR) may be used to increase the size and broaden the size distribution of the seed crystal population; it can also provide graded surface coverage of the crystals and modify their morphology. This chapter provides tools to achieve these aims using designed experiments, and to control growth and crystal size distribution of a seed crystal population when no spontaneous nucleation (renucleation) takes place. The growth of crystals in seeded CSTR crystallizers is modeled as a function of the addition rate of the seed crystals, seed crystal size and number, morphology, growth rate, and residence time.

The modeling focuses on predicting the reactant addition rate necessary for the controlled growth of the seed crystal population as determined by these parameters. To obtain constant growth rates of the crystal growth in the seeded CSTR crystallizer, the results predict that the reactant addition rate must be nonlinearly adjusted for seed size, growth rate, and residence time, and linearly for seed number. The modeling thus gives information for the control of the growth process in seeded CSTR crystallizers. The determination of the addition rate by the present model gives the experimenter a tool to precisely predetermine the size and size distribution of the crystal population in a seeded growth CSTR crystallizer at steady state.

INTRODUCTION

Crystallization in continuous stirred tank reactor (CSTR, or mixed-suspension mixed-product-removal, MSMPR) crystallizers is important for the production of crystalline materials.[1] This process has generally been discussed under the assumption of spontaneous nucleation. In the case of spontaneous nucleation, the competition of growth and nucleation determines the growth rate of the crystal population. The BNG model, as discussed in the previous chapters, correlates the maximum crystal size with the reaction conditions. In contrast, the Randolph-Larson (R-L) model focuses on the dispersity of the crystal population as determined by the growth rate and the residence time. The growth rate is determined from the crystal size distribution at steady state and the residence time. A nucleation rate is determined by extrapolation to zero crystal size.[1,2,3]

The process of seeded CSTR crystallization is a variation of the CSTR crystallization in which spontaneous nucleation is replaced by controlled addition of (seed) crystals. If spontaneous nucleation is suppressed, the seed crystals will grow during

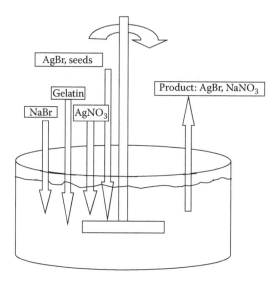

FIGURE 10.1   Schematic of seeded CSTR reactor: AgBr precipitation.

steady state at constant growth rates. The growth rate of the seed crystals below the maximum growth rate is determined by seed addition rate, seed size and morphology, reactant addition rate, and residence time. Due to the dispersion of residence time of the crystals in the reactor, the crystal size distribution will widen.

It appeared that no model had been developed to explicitly relate the growth rate in the seeded CSTR crystallization to the reactant addition rate. Thus, a model was developed where the growth rate was explicitly related to seed addition rate, seed size and morphology, reactant addition rate, and residence time. The knowledge of the growth rate and the residence time allows predicting the crystal size distribution at steady state.[4]

An example for a seeded CSTR (MSMPR) reactor is shown in Figure 10.1 for the continuous growth of silver bromide, AgBr. The reactants, silver nitrate ($AgNO_3$) and sodium bromide (NaBr), the AgBr crystal seeds, and an aqueous solution of gelatin are added at controlled flow rates to the reactor. To control the residence time, product removal is at the same flow rate as the sum of input flow rates. The procedure will be discussed in greater detail under experimental considerations. Without adding AgBr seed crystals, the reactor will spontaneously form silver bromide crystals.[5]

Before presenting the proposed model for the seeded CSTR crystallizer, it may be useful to provide a background for the crystallization in continuous stirred tank reactor (CSTR) crystallizers with spontaneous nucleation.

Two independent models have been proposed by Randolph and Larson[1] and Leubner.[6] The model by Randolph and Larson (R-L model) is represented by Equation (10.1).[1]

$$n_x = n_0 \exp\left(-\frac{L_x}{G\tau}\right) \tag{10.1}$$

Here, n_x is the number population density of crystals (number/s cm^3) with the size L_x, n_0 is the population density of newly formed crystals (nuclei, nascent crystals) at steady state (birth rate, B^0), G is the growth rate (dL_x/dt), and τ is the residence time of the reactor.

Nucleation rate, B^0, and growth, G, are proposed to be linked by a power relationship as shown in Equation (10.2). The exponent, i, determines the stability of the steady state, where $i > 21$ is predicted to cause instability.[1]

$$B^0 \sim G^i \qquad (10.2)$$

The residence time, τ, is defined as the reaction volume divided by the sum of the incoming (or outgoing) reaction streams. In this model, the growth rate, G, is determined by evaluating the slope of a plot of $\ln (n_x/n_0)$ versus L_x. It is assumed that the number density (number/cm^3) is a rate-determining factor. In Equation (10.1), the nucleation rate, n_0, is determined by extrapolation to $L_x =$ zero. The growth rate is the result of the mass balance between reaction addition, nucleation, and growth. The R-L model does not explicitly relate the nucleation or growth rate to critical reaction conditions such as temperature, solubility, or reactant addition rate. In the previous chapters we presented an independent and different model based on the BNG model for the crystallization in the CSTR crystallizer for systems with homogeneous nucleation.[6] This model is an extension of the balanced nucleation and growth (BNG) model developed for semibatch precipitations.[7] Equation (10.3) describes the average crystal size, L, as a function of the reaction conditions.

$$k_v L^3 R_g T - 2k_s \gamma D V_m^2 C_s \Psi - 3k_v G_m R_g T L^2 \tau = 0 \qquad (10.3)$$

Here, k_v and k_s are volume and surface shape factors that convert size to crystal volume and surface, respectively; L is the average crystal size of the crystal population at steady state; R_g is the gas constant; T (K) is the reaction temperature; γ is the crystal surface energy; D is the diffusion constant of the reactants; V_m is the molar volume (cm^3/mole of the crystal composition); C_s is the solubility of the crystals; Ψ is equal to ($L/L^* -1.0$) where L is the average crystal size; L^* is the (smaller) critical crystal size at which crystals have equal probability to grow or dissolve in the reaction system; G_m is the maximum growth rate of the crystals; and τ is the residence time as defined under Equation (10.1). Solving Equation (10.3) for L leads to complicated third-order equations that are difficult to evaluate. As shown in the previous chapters, it is preferentially solved for the reaction variables residence time, τ, and solubility, C_s, which results in relatively straightforward equations that are more transparently solved.

Both the R-L and the BNG model assume spontaneous crystal nucleation.[1,4] The present work is a continuation of the modeling of the crystallization in the CSTR system where spontaneous nucleation is replaced by designed continuous addition of seed crystals. For steady-state conditions, the crystal size distribution and growth rate will be related to the seeding conditions (size, morphology, number, and reaction addition rate), residence time, and reactant addition rate. It will be assumed that growth of the seed crystals will fully repress the spontaneous formation of new nuclei (renucleation).[8]

SEEDED CRYSTALLIZATION IN THE CSTR SYSTEM

Controlled crystallization in a CSTR crystallizer is a complex process in which a variety of variables must be closely controlled. An example is the continuous precipitation of silver halides as discussed in the previous chapters. The crystallization in the seeded CSTR crystallizer adds the controlled addition of the seed crystals and the variation of seed addition rate, seed size, and seed morphology.

In precipitations with spontaneous nucleation, continuous streams of reactants (silver nitrate and halide salt solutions) and an aqueous gelatin solution are fed into a well-stirred precipitation vessel, and product is simultaneously removed. During the precipitation, a constant reaction volume and constant reaction conditions (temperature, solubility, and residence time) are maintained. The gelatin is a protective colloid that prevents agglomeration of the silver halide crystals. Silver halides have very low solubility, and the output stream consists of an aqueous solution of silver halide crystals, gelatin, and nitrate salts. The silver halide crystals have polydisperse size distribution but homogeneous morphology. The size is relatively symmetrically distributed around an average crystal size.

In the present experimental model, an additional input stream is added that contains a monodisperse population of crystals with well-defined size and morphology, which are referred to as "seed crystals" or "seeds." It is assumed that the seed crystals will consume all incoming reactant material for growth. Their dissolution in the reaction mixture is suppressed by the supersaturation due to added growth material. The presence of the seed crystals will suppress spontaneous nucleation (renucleation) until the reactant addition rate exceeds the maximum growth rate, G_m, of the crystal population. Renucleation was discussed, studied, and modeled for semibatch precipitations in Chapter 8 and may be applied when spontaneous nucleation adds to the growth of the seeds.

The experimental procedure for the study of seeded CSTR crystallizations is thus relatively straightforward. However, for the controlled precipitation and growth of crystals in the CSTR crystallizer, certain conditions are required:

1. Reactor conditions, temperature, solubility of the crystalline material, and residence time must be closely controlled and held constant for the time of the experiment.
2. Tight control of the residence time by control of input and output flow rates achieves constant reaction volume.
3. Flow rates and reactant concentrations are tightly controlled to keep the suspension density constant and to provide controlled growth of the crystal population.
4. Steady state of mass balance is reached in four to five residence times after the beginning of the reaction if the reaction volume is correctly adjusted at the start of the precipitation.
5. However, steady state of growth and nucleation for unseeded crystallizations was obtained as late as eight to 12 residence times after the start of the precipitation, which is significantly later than that of mass balance.[5]

6. For *seeded crystallizations*, the number of added crystals, their size, and morphology must remain constant throughout the precipitation. This is in addition to tight control of the flow rate of the solution that contains the seed crystals. For practical purposes, the seeds may be contained in one of the other reaction streams, preferentially the one without reactants, for instance, in the case of silver halide precipitation in the stream containing gelatin. The high silver ion and halide concentrations in the reactant streams would dissolve seed crystals and result in uncontrolled seeding of the reactor (Figure 10.1).

7. At steady state of mass balance, nucleation, and growth, there is a steady state of incoming seed crystals, and a steady-state distribution of growing crystals.

8. For the modeling of the seeded CSTR crystallizer, the R-L model will be used as the starting base.

9. To avoid multiple reaction zones, the shape of the reactor, stirring, and mixing must be controlled. Multiple reaction zones result in uncontrolled cycling of the system and thus non-steady-state reaction conditions, leading to uncontrolled dispersity of the product as a function of time.

MODELING THE SEEDED CSTR CRYSTALLIZER

At steady state, the number of incoming seed crystals (N_0) that have not grown is given by their addition rate (crystal number/s, N/s). This is different from the number population, n_0, in the Randolph-Larson model (Equation [10.1]), which is normalized to time and reaction volume (N_0/s cm³). The number of crystals left after the time, t ($= N_x$), is given by Equation (10.4).

$$N_t = N_0 \exp \frac{-t}{\tau} \qquad (10.4)$$

Here, τ is the residence time. In Figure 10.2, the decrease of crystal number after seeding is shown for residence times of 1, 5, and 10τ. The plots for τ equal to 1 and 5, showing that the "washout" of the seed population is complete after about five residence times if no new crystals are added.

During the time, t, the crystal population, N_x, has grown from the initial size L_0 to size L_x:

$$L_x = L_0 + Gt \qquad (10.5)$$

This assumes size- and time-independent growth rate. If there is a known size or time dependence of the growth rate, this can be incorporated in Equation (10.5). The growth of a crystal is a function of the time that it resides in the reactor. The growth between addition of the crystals and their removal from the reactor thus provides a defined increase in crystal size. The time of growth is a function of the reactor residence time, τ, as given in Equation (10.4).

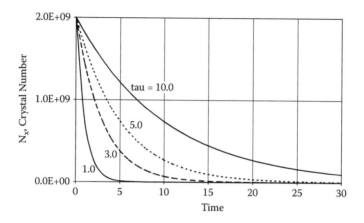

FIGURE 10.2 Decrease of seed crystal number, N_x, with time for $\tau = 1$, 5, and 10. Initial seed number $N_0 = 2.0 \times 10^9$.

If Equation (10.5) is solved for t and inserted into Equation (10.4), then Equation (10.6) is obtained, which is a modified version of Equation (10.1) for seeded CSTR precipitations at steady state.

$$N_x = N_0 \exp \frac{L_0 - L_x}{G\tau} \qquad (10.6)$$

Equation (10.6) can be experimentally evaluated by plotting log N_x against L_x.

In Figure 10.3, this is modeled for residence times of 1, 3, 5, and 10τ. A seed number of 20×10^9, seed size of 0.20, and growth rate of 0.050 were used.

The growth of the seed crystals is also determined by the addition rate of the reactants. For monodisperse crystal systems in semibatch precipitations, the relationship

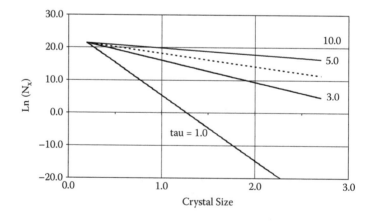

FIGURE 10.3 Size distribution, Ln(N_x), versus crystal size for $\tau = 1$, 3, 5, and 10 for an initial seed number of 20×10^9, seed size 0.20, and growth rate 0.050.

between growth rate, G, and molar addition rate, R (mol/s), of the reactants is given by the mass-balance equation, (10.6), where N is the number of crystals of size L in the reactor.

$$G = \frac{dL}{dt} = \frac{RV_m}{3.0k_v L^2 N} \tag{10.7}$$

With the assumption that the growth rate is size independent, Equation (10.6) can be rewritten at steady state for each size, L_x, in the seeded CSTR reactor (Equation [10.8]).

$$G = \frac{R_x V_m}{3.0k_v L_x^2 N_x} \tag{10.8}$$

Here, R_x is the fraction of the incoming reactant stream (R, mole/s) consumed for growth by the crystal population fraction, N_x, with the size, L_x. For further modeling, Equation (10.8) is solved for R_x (Equation [10.9]).

$$R_x = \frac{3.0k_v GL_x^2 N_x}{V_m} \tag{10.9}$$

The total addition rate is the sum of the addition rates, R_x, for the crystal size range, $L_x = L_0$ to $L_x = \infty$.

$$R = \int_{L_0}^{\infty} R_x \, dL \tag{10.10}$$

Equation (10.10) can be expanded by back substitution of R_x from Equation (10.9).

$$R = \frac{3.0k_v G}{V_m} \int_{L_0}^{\infty} L_x^2 N_x \, dL \tag{10.11}$$

Back substitution of Equations (10.4) and (10.5) and replacing dL with dt leads to Equation (10.12).

$$R = \frac{3.0k_v N_0 G}{V_m} \int_{t=0}^{\infty} (L_0 + Gt)^2 \exp\left(\frac{Gt - L_0}{G\tau}\right) dt \tag{10.12}$$

For the present work, Equations (10.4), (10.5), (10.9), and (10.10) were used for stepwise numerical calculations using a spreadsheet. Alternatively, Equation (10.12) may be numerically integrated (Equation [10.13]).

$$R = \frac{3.0k_v N_0 G}{V_m} \left(\sum_{i=0}^{i=\infty} (L_0 + G(t_0 + i \ t)^2 \exp\left(\frac{G(t_0 + i \ t) - L_0}{G\tau}\right) \right) \tag{10.13}$$

The starting time for the calculations is conveniently defined by t_0 equal to zero. These equations allow the calculation of the addition rate that must be provided to sustain the desired growth conditions of the seed crystals. To avoid renucleation, G must be less than the maximum growth rate, G_m. Alternatively, G_m may be determined by varying the reaction rate (R, mole/s), seed addition rate (number/s), or residence time (s), and experimentally determining the condition under which renucleation occurs.

MATHEMATICAL PROCESSING STEPS

For the calculations for steady-state conditions, certain experimental variables must be given: the seeding rate of the added crystals, N/dt; the size, L_0, of the seed crystals; volume shape factor, k_v, which is determined by the morphology of the seed crystals; and the residence time, τ. For the current calculations, cubic morphology is assumed for which k_v is equal to 1.0. Adjustable variables for the calculations are the residence time, τ, the growth rate, G (dL/dt), seed size, and seed addition rate (dN/dt).

It is the aim to calculate the total addition rate, R (mole/s), needed to obtain a desired growth rate of the crystal population as a function of seed addition rate, size, morphology, and residence time. In this process, the model also predicts the size distribution, N_x, versus L_x at steady state and the uptake rate R_x for each size class defined by L_x and N_x.

The following sequence of calculations was performed using a standard spreadsheet:

1. A time range, t, a residence time, τ, and time intervals, dt/τ, were chosen. To provide steady-state conditions, t was generally chosen equal to 10τ.
2. N_x values were calculated as a function of t (Equation [10.4]). This determines the number of seed crystals remaining in the reactor after the time, t, of seed addition.
3. L_x values were calculated for the corresponding N_x (Equation [10.5]) using t and the growth rate, G.
4. R_x values, the fraction of the reactant input rate used, N_x, were calculated for the L_x (N_x) crystal population (Equation [10.9]).
5. The total reactant addition rate R needed to provide the desired growth rate, G, for the seed crystals was calculated using Equation (10.13).
6. The input parameters were changed and the calculations were repeated for the new input conditions. The results of the calculations were used to plot the curves in the figures.

The determination of the addition rate using the present model gives the experimenter a tool to predetermine precisely the size and size distribution of the crystal population at steady state.

RESULTS AND DISCUSSION

The concepts and equations derived were used to model the dependence of crystal size and size distribution in a seeded CSTR crystallizer as a function of the critical reaction variables seed number, N_0, seed size, L_0, growth rate, G, and residence time, τ.

TABLE 10.1
Ranges of Reaction Variables: Seed Size, L_0,
Growth Rate, G, and Residence Times, τ

Seed size, L_0	0.05–1.0
Growth rate, G	0.001–1.0
Residence time, τ	1.0–10.0
Seed number	0.5–9.0×10^9
Fixed variables: $V_m = 25.0$, $k_v = 1.0$	

For the initial conditions of the reaction, it is assumed that all reaction variables are at their steady-state conditions and that the reactor is initially not seeded. The calculated addition rates in the figures were determined for the time range zero to 10τ to assure that they represent steady state. Relative addition rates (% addition rate) were calculated with the 10τ value as the 100% steady-state reference. The constants, variables, and their ranges used for the calculations are compiled in Table 10.1.

Addition Rate, R, as a Function of Residence Time, τ, for Constant Growth Rate, G

In Figure 10.4, the relative molar addition rate (% addition rate, relative to addition rate at $t = 10\tau$) to achieve constant growth rate is shown as a function of time for the initial phase of the seeding/growth. The residence times were varied from 1 to 10τ.

Figure 10.4 shows that the reactant addition rate must be continuously varied during the initial stage of the reaction to assure constant crystal growth rate. Thus, the control of the addition rate during the initial transition phase has a significant effect on the establishment of the steady state. Initial addition rates that are higher than necessary may lead to renucleation, which will significantly modify the establishment of the steady state of seed addition and growth.

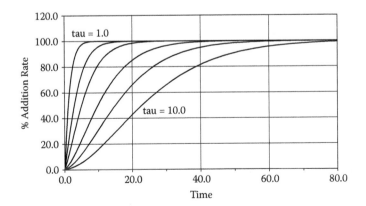

FIGURE 10.4   Percent addition rate versus time for τ = 1, 2, 3, 5, and 10. $N_0 = 2.0 \times 10^9$, $L_0 = 0.200$, G = 0.050.

TABLE 10.2

Steady-State Addition Rate as a Function of Residence Time

Residence Time, τ	Addition Rate, R
1	1.30E+04
2	4.00E+04
3	8.19E+04
5	2.64E+05
7	5.44E+05
8	8.30E+05
9	1.12E+06
10	1.34E+06

$N_0 = 2.0 \times 10^9$, $L_0 = 0.200$, $G = 0.050$, at $t = 10\tau$

Figure 10.4 also shows that about eight to 10 residence times are needed to achieve constant addition rate for steady-state conditions for crystal growth. This is about twice the time to achieve steady state of the mass balance (washout condition, Equation [10.4], Figure 10.2). With this in mind, the dependence of addition rate at steady state will be reported for the variation of the model variables.

Steady-State Addition Rate as a Function of Growth Rate, G

Addition rates at steady state vary over several orders of magnitude as a function of growth rate (Table 10.2). Thus, the addition rates are plotted logarithmically versus growth rate in Figure 10.5. It shows that the addition rate to sustain steady growth rate increases strongly at low growth rates. At high growth rates, the dependence of addition rate on growth rate becomes significantly less.

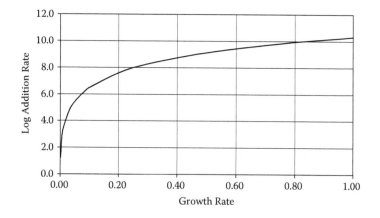

FIGURE 10.5 Log addition rate at 10τ (steady state) versus growth rate. $N_0 = 2.0 \times 10^9$, $L_0 = 0.200$, $\tau = 5.0$.

TABLE 10.3

Steady-State Addition Rate as a Function of Growth Rate

Growth Rate, G	Addition Rate, R
0.001	1.61E+01
0.005	5.13E+02
0.010	2.60E+03
0.030	5.21E+04
0.050	2.64E+05
0.090	2.02E+06
0.100	2.95E+06
0.200	3.89E+07
0.300	1.85E+08
0.500	1.35E+09
0.750	6.66E+09
1.000	2.08E+10

$N_0 = 2.0 \times 10^9$, $L_0 = 0.200$, $\tau = 5.0$, at $t = 10.0\tau$

The modeling of the addition rate as a function of aim growth rate gives quantitative guidance to control the addition rate during the start-up time of the seeded CSTR crystallizer.

Steady-State Addition Rate, R, as Function of Residence Time, τ

As the residence time is varied, the reactant addition rate also must be varied to achieve a given growth rate. The model was used to calculate the addition rates as a function of residence time to sustain a constant growth rate (Table 10.3). For Figure 10.6, the addition rate necessary to sustain constant growth rate at time

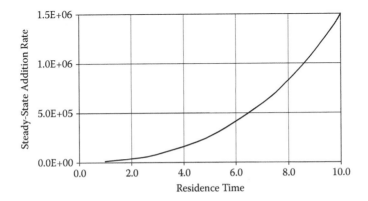

FIGURE 10.6 Addition rate versus residence time at 10τ (steady state). $N_0 = 2.0 \times 10^9$, $L_0 = 0.200$, $G = 0.050$.

TABLE 10.4

Steady-State Addition Rate as a Function of Seed Size

Seed Size, L_0	Addition Rate, R
0.05	1.52E+05
0.10	1.84E+05
0.15	2.22E+05
0.20	2.64E+05
0.30	3.64E+05
0.40	4.84E+05
0.50	6.25E+05
0.70	9.65E+05
1.00	1.63E+06

$N_0 = 2.0 \times 10^9$, $L_0 = 0.200$, $G = 0.050$, $\tau = 5.0$, at $t = 10\tau$

10τ was plotted versus residence time. As shown in Figure 10.4, steady state is expected to be reached at $t = 80$ for 10 τ. Figure 10.6 shows that the addition rate must be nonlinearly increased with increasing residence time to sustain a constant growth rate.

Steady-State Addition Rate, R, as Function of Seed Size, L_0

It is anticipated that the addition rate must be varied if the size of the seed crystals is varied (Table 10.4). The results of this modeling show that the addition rate must be nonlinearly increased with seed size to sustain constant growth rate when all other variables are held constant (Figure 10.7).

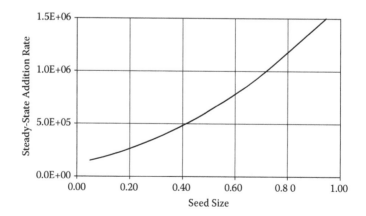

FIGURE 10.7 Addition rate at 10τ (steady state) versus seed size. $N_0 = 2.0 \times 10^9$, $G = 0.050$, $\tau = 5.0$.

TABLE 10.5
Steady-State Addition Rate as a
Function of Seed Number

Seed Number, $*10^9$	Addition Rate, R
0.50	6.61E+04
1.00	1.32E+05
2.00	2.64E+05
4.00	5.29E+08
6.00	7.93E+05
7.00	9.25E+05
8.00	1.06E+06
9.00	1.19E+06

$L_0 = 0.200$, $G = 0.050$, $\tau = 5.0$, at $t = 10\tau$

Steady-State Addition Rate, R, as a Function
of Seed Number Addition Rate, dN_0/dt

If the addition rate of the crystals, dN_0/dt, is varied, then the reaction addition rate must be varied if all other factors are to be kept constant (Table 10.5). In Figure 10.8, the results modeling this condition are shown. Figure 10.8 shows that the addition rate must be linearly increased with seed number addition rate when seed size, residence time, and growth rate are held constant. The linear correlation makes it straightforward to predict the reactant addition rate as a function of seed addition rate.

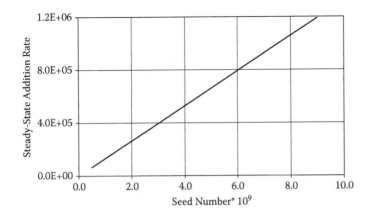

FIGURE 10.8 Addition rate at 10τ (steady state) versus seed number. $L_0 = 0.200$, $G = 0.050$, $\tau = 5.0$.

CONCLUSION

The growth of crystals in a continuous precipitation system is different from the previous models by Leubner (BNG model) and Randolph-Larson (R-L model), since the introduction of crystal nuclei is experimentally controlled by seed addition rate rather than by spontaneous nucleation. To control the growth and crystal size dispersity in the seeded system, the growth rate of the crystals is controlled by the reactant addition rate, the rate of seed addition, the morphology of the seed crystals, and the residence time. The present work provides a model by which the precision control of the seed crystal growth and size dispersity is controlled. It provides equations that are essential to adjust the reaction parameters to obtain the desired results for product crystal size and size distribution.

The balanced growth model for the seeded CSTR crystallizer assumes that the growth rate is based on the mass balance between reactant addition rate and growth of the available crystal surface. It is further assumed that spontaneous nucleation is suppressed by the growth processes of the added seed population. Mixing is assumed to lead to a homogeneous reaction system. The reaction is not dependent on the volume of the reactor but on the residence time, which is a function of reaction volume and volume addition rate or equivalently by the volume removal rate. Thus, the modeling of the reaction need not consider the reaction volume at steady-state conditions.

The present model for seeded growth in the CSTR crystallizer predicts that for constant growth rate, the addition rate must be increased during the initial phase of the reaction to match the changing crystal population. The model also predicts that steady state of growth and seed addition is reached after eight to 10 residence times when the reactor was not precharged with seed crystals.

The reactant addition rate for balanced growth of seed crystals in CSTR crystallization is modeled as a function of the number addition rate of seeds, their size, morphology, growth rate, and the residence time of the reactor. The reactant addition rate necessary to obtain a given growth rate of the seed crystal population is predicted from these parameters.

The modeling shows that the reactant addition rate must be nonlinearly increased for increasing seed size, growth rate, and residence time, and it must be linearly increased for increasing seed number addition rate to control the crystal growth in a seeded CSTR crystallization at a given growth rate.

The crystal size distribution is a function of the product, $G \times \tau$, that is, growth rate and residence time. Here, τ is a control variable, and G can be controlled by the application of the presented model. The results of the balanced growth model thus provide quantitative information to control the growth process and size distribution in seeded CSTR crystallizers.

REFERENCES

1. Randolph, A. D., and M. A. Larson. 1991. *Theory of particulate processes, analysis, and techniques of continuous crystallization*. 2nd ed. San Diego: Academic Press.
2. Bransom, H. S., W. J. Dunning, and B. Millard. 1949. *Disc Faraday Soc* 5:83.
3. Randolph, A. D., and M. A. Larson. 1962. *AIChE J* 8:639.

4. Leubner, I. H. 2001. Seeded precipitations in the continuous stirred tank reactor crystal-lizer. *J Disp Sci Technol* 22:373–80.
5. Wey, J. S., I. H. Leubner, and J. P. Terwilliger. 1983. Transient behavior of silver bro-mide precipitation in a continuous suspension crystallizer. *Photogr Sci Eng* 27:35.
6. Leubner, I. H. 1998. A new crystal nucleation theory for continuous precipitation of silver halides. *J Imaging Sci Technol* 42:355.
7. Leubner, I. H. 2000. Particle nucleation and growth models. *Curr Opinions Colloid Interface Sci* 5:2.
8. Leubner, I. H. 1993. Crystal growth and renucleation, theory and experiments. *J Imaging Sci Technol* 37:510.

Epilogue

FINAL THOUGHTS

Thank you for reading this far. I hope that you learned practical concepts to control your crystallization processes. You have struggled through the nucleation phase, batch and continuous processes, addition rate, solubility, temperature, residence time, nano- and super-sizing with restrainers and ripeners, renucleation, and others. May the information and models stimulate your thoughts on how to apply these control variables to your system. May they guide you to reduce the number of experiments—to as low as two or three—to define your system, leading you to quantitatively predict the experimental parameters that will lead to the desired crystals. Surprising predictions were experimentally confirmed and erroneous concepts were corrected. Best of all, the BNG model, as described here, may help you shorten the time needed for research and product development, and replace time-consuming trial and error procedures. It is my desire that this work may stimulate new experiments, research, and the study and teaching of crystallization. It is my hope that this book may become an important part of your crystallization library.

Ingo H. Leubner

Glossary of Terms

Variable	Definition	Comment
C	Concentration	mol/l
C_0	Solubility for L = 0 (extrapolated)	mol/l
cel	Cubic edge length	μm
C_m	Minimum solubility	
C_s	Solubility	mol/cm^3
C_{ss}	Supersaturation	mol/cm^3
CSTR	Continuous stirred tank reactor	
D	Diffusion constant	cm^2/s
ecd	Equivalent circular diameter	μm
F	Flow rate	ml/min
F_t	Total flow rate (l/min)	l/min
G	Crystal growth rate	dL/dt, dr/dt
G_m	Maximum crystal growth rate	cm/s, nm/s, A/s
k_s	Crystal surface constant	
k_v	Crystal volume constant	Cubic: 1.0
l	Liter	
L	Average crystal size	cm, μm
L*	Critical crystal size	cm, μm
L_0	Initial size of seed crystals	
L_c	Critical crystal size	cm, μm
L_n	Nascent crystal size	cm, μm
L_x	Seed crystal size at time t, Crystal size (R-L model)	cm
ml	Milliliter	
MSMPR	Mixed-suspension mixed-product-removal reactor	
M_t	Suspension density	mol/l
N	population density, number/(volume-length)	No/cm × cm^3
n^0	nuclei population density, number/(volume-length)	n/s cm3
N_0	Seed addition rate, crystal number/s	N/s
n_x	Number population density of crystals (n/s cm^3)	n/cm × cm^3
N_x	Seed crystal number at time t	
pAg	-log [Ag$^+$]	
r	Radius of crystal	
R	Reactant addition rate (mole/s)	mol/s
R_0	Addition rate of reactants (CSTR)	mol/s
R_g	Gas constant	erg/deg mol
R_i	Addition rate fraction used for growth	
R_n	Addition rate fraction used for nucleation	
R_x	Fraction reactant addition rate (mole/s)	

(continued)

Variable	Definition	Comment
S	Supersaturation	mol/L
S*	Supersaturation ratio	
S_m	Characteristic surface (e.g., area/mol silver halide)	
T	Temperature	K, °C
t	Reaction time, s	
V	Reaction volume	mL
V_0	Initial reaction volume	ml
V_g	Average crystal volume	cm^3
V_m	Molar volume	cm^3/mol, crystals
Y	$L/L^* - 1.0$	
Z	Number of crystals	
Z_n	Number of crystals nucleated at steady state	
Z_r	Number of crystals in the reactor during steady state	crystals/ml
Z_t	Total number of crystals in reactor during steady state	
γ	Crystal surface energy	erg/cm^2
τ	Residence time	Min, sec

Bibliography

Berriman, R. W. *J Photogr Sci* 12 (1964): 121.

Berry, C. R., and D. C. Skillman. *J Phys Chem* 68 (1964): 1138.

Bransom, H. S., W. J. Dunning, and B. Millard. *Disc Faraday Soc* 5 (1949): 83.

Brito, J., and R. H. Heist. *Chem Eng Commun* 15 (1982): 133.

Chateau, H., J. Pouradier, and C. R. Berry. In *The Theory of the Photographic Process*. 3rd ed. Edited by T. H. James. New York: Macmillan, 1971.

Dann, J. R., P. P. Chiesa, and J. W. Gates. *J Org Chem* 26 (1961): 1991.

Daubendiek, R. L. *Proc Int Congr Photogr Sci* (1978): 141.

Frensdorf, H. *J Am Chem Soc* 27 (1971): 1.

Greeley, W., et al. *J Am Chem Soc* (1960), 64, 652.

Gutoff, E. B. *Photogr Sci Eng* 14 (1970): 248.

Gutoff, E. B. *Photogr Sci Eng* 15 (1971): 189.

Heist, R. H., and A. Kacker. *J Chem Phys* 82 (1985): 2734.

Heist, R. H., A. Kacker, and J. Brito. *Chem Eng Commun* 28 (1984): 117.

Hengel, R. *Photogr Sci Eng* 27 (1983): 1.

Hurle, D. T. F., ed. *Handbook of Crystal Growth*. Elsevier, 1993.

Jagannathan, R. *J Imaging Sci* 32 (1988): 100.

James, T. H. Chap. 1 in *The Theory of the Photographic Process*. 4th ed. Edited by T. H. James. New York: Macmillan, 1977.

Katz, J. L., and M. D. Donohue. *Adv Chem Phys* 40 (1979): 137.

Kharitanova, A. I., B. I. Shapiro, and K. S. Bogomolov. *Z Nauchn Prikl Fotogr Kinematogr* 24 (1979): 34.

King, T. W., S. M. Shor, and D. A. Pitt. *Photogr Sci Eng* 25 (1981): 70.

Klein E., and E. Moisar. *Ber Bunsenges Phys Chem* 67 (1963): 349.

Kresta, S. M., G. L. Anthieren, and K. Parsiegla. *Chem Eng Res Design* 82, no. A9 (2004): 1117–36.

Leubner, I. H. Balanced nucleation and growth model for controlled crystal size distribution. *J Disp Sci Technol* 23 (2002): 577.

Leubner, I. H. A balanced nucleation and growth model for controlled precipitations. *J Disp Sci Technol* 22 (2001): 125–38.

Leubner, I. H. *Colloids and Surfaces in Reprographic Technology*. ACS Symposium Series 22, 1982.

Leubner, I. H. Crystal formation (nucleation) in the presence of growth restrainers. *J Crystal Growth* 84 (1987): 496.

Leubner, I. H. Crystal formation (nucleation) in the presence of Ostwald ripening agents. *J Imaging Sci* 31, no. 4 (1987): 145–48.

Leubner, I. H. Crystal formation (nucleation) of silver halides. Comparison of models. 45th Annual Conference of the Society for Imaging Science and Technology, East Rutherford, NJ.

Leubner, I. H. Crystal formation (nucleation) under kinetically controlled and diffusion-controlled growth conditions. *J Phys Chem* 91 (1987): 6069–73.

Leubner, I. H. Crystal growth and renucleation: theory and experiments. *J Imaging Sci Technol* 37 (1993): 510.

Leubner, I. H. Formation of silver halide crystals in double-jet precipitations: AgCl. *J Imaging Sci* 29 (1985): 219–25.

Leubner, I. H. A new crystal nucleation theory for continuous precipitation of silver halides. *J Imaging Sci Technol* 42 (1998): 355.

Leubner, I. H. Number and size of AgBr crystals as a function of addition rate: A theoretical and experimental review. *J Imaging Sci Technol* 37 (1993): 68.

Leubner, I. H. Particle nucleation and growth models. *Curr Opin Colloid Interface Sci* 5 (2000): 2.

Leubner, I. H. Seeded precipitations in the continuous stirred tank reactor crystallizer. *J Disp Sci Technol* 22 (2001): 373–80.

Leubner, I. H., R. Jagannathan, and J. S. Wey. Formation of silver bromide crystals in double-jet precipitation. *Photogr Sci Eng* 24 (1980): 268–72.

Lide, D. R., ed. *Handbook of Chemistry and Physics*. 85th ed. Boca Raton, FL: CRC Press, 2004.

Ludwig, F. P., and J. Schmelzer. *Zeitschrift für Physikalische Chemie* 192 (1995): 155–67.

McBride, C. E. U. S. Patent 3,271,157, 1966.

Mersman, A., ed. *Crystallization Technology Handbook*. 2nd ed. Marcel Dekker, 2001.

Muhr, H. R., J. Villermaux, and P. H. Jezequel. *Chem Eng Sci* 50 (1995): 345–55.

Muhr, H. R., J. Villermaux, and P. H. Jezequel. *Chem Eng Sci* 51 (1996): 309–19.

Mullin, J. W. *Crystallization*. 4th ed. Oxford: Butterworth-Heinemann, 2000.

Mutaftschiev, B. *Handbook of Crystal Growth*. Vol. 1a. Edited by D. T. F. Hurle. Amsterdam: North-Holland, 1993.

Myerson, A. S. *Handbook of Industrial Crystallization*. Butterworth-Heinemann, 1993.

Pouradier J., A. Pailliotet, and C. R. Berry. in *The Theory of the Photographic Process*. 4th ed. Edited by T. H. James. New York: Macmillan, 1977.

Randolph, A. D., and M. A. Larson. *Theory of Particulate Processes, Analysis and Techniques of Continuous Crystallization*. Academic Press, Second edition, 1991.

Randolph, A. D., and M. A. Larson. *AIChE J* 8 (1962): 639.

Strong, R. W., and J. S. Wey. *Photogr Sci Eng* 23 (1979): 344.

Sugimoto, T. Proceedings of the 11th Symposium on Industrial Crystallization, Garmisch-Partenkirchen, Germany, 1990.

Tanaka, T., and M. Iwasaki. *J Imaging Sci* 29 (1985): 20.

Tavare, N. S. *AIChE J* 33 (1987): 152.

Towns, M. B., et al. *J Phys Chem* 64 (1960): 1861.

Wagner, C. Z. *Elektrochemie* 65 (1961): 581.

Wey, J. S., and R. Jagannathan. *AIChE J* 28 (1982): 697.

Wey, J. S., and R. W. Strong. *Photogr Sci Eng* 21 (1977): 14.

Wey, J. S., and J. P. Terwilliger. *AIChE J* 20 (1974): 1219.

Wey, J. S., I. H. Leubner, and J. P. Terwilliger. *Photogr Sci Eng* 27 (1983): 35.

Wey, J. S., J. P. Terwilliger, and A. D. Gingello. *Res Discl* (1976): 14987.

Wey, J. S., J. P. Terwilliger, and A. D. Gingello. *Analysis of AgBr Precipitation in a Continuous Suspension Crystallizer*. AIChE Symposium Series 193, 1980.

Index

"f" indicates material in figures. "t" indicates material in tables.

A

1-(3-acetamidophenyl)-mercaptotetrazole, 91,
  96–100, 97t
Addition rate
  in Becker-Doering model, 5
  in BNG model, 29
    assumptions for, 11, 18
    in continuous nucleation, 135–142, 154
    crystal number and, 18, 21
    crystal size and, 21, 21f
    growth and, 16
    in nucleation phase, 13, 14, 20–21
    time and, 20f
    value range for, 19t
  concentration and, 125, 136
  in continuous nucleation
    BNG model for, 135–142, 154
    crystal size and, 135, 143, 147, 152, 153
    growth and, 125, 135, 141, 143, 152, 154
    in initial transient phase model, 124–126
    residence time and, 144t, 152
    in R-L model, 119, 121
    suspension density and, 145t
    yield and, 117
  convection rate and, 7
  critical/average size ratio and, 47t, 50f, 51,
    52f
  crystal number and; See under Crystal
    number, addition rate and
  crystal size and; See under Crystal size,
    addition rate and
  CSD and, 21, 21f, 24–25
  diffusion rate and, 7
  in diffusion-controlled nucleation, 30, 32,
    37–38, 42, 44–53, 137
  for double-jet precipitations, 7
  edge length and, 125
  flow rate and, 125, 136
  growth and; See under Growth, addition
    rate and
  in kinetically-controlled nucleation, 31–34, 94
  in Klein-Moisar model, 5–6
  in mass-balance model, 136
  molar volume and, 13, 125
  nucleation rate and, 20f, 21
  nuclei volume and, 15
  in primitive model, 1–2

reaction rate and, 7
reaction volume and, 125
renucleation and, 101–113, 109t, 113f, 175
of restrainers, 97
ripeners and, 76
in R-L model, 119, 121
for seeded CSTR
  discussion of, 167–179
  growth and, 173–177, 176f, 177t
  residence time and, 176t, 177f
  seed number and, 179f, 179t
  seed size and, 178f, 178t
  time and, 175f
for silver bromide, 45, 47t, 51–53, 52f
for silver chloride, 47t, 48–50, 50f
solubility and, 7
at steady state, 138
supersaturation and, 14, 14t, 44
surface area and, 125
suspension density and, 124–125, 145t
temperature and, 7
Adenine, 91–92, 95
Adsorption, 26, 83, 88t, 91–96
Agglomeration, 48–49, 170
Ammonia, 69, 77–83, 79f, 80f, 81f
Area-weighted crystal size, 31
Arrhenius reaction rate, 4–5, 13
Attenuation model, 122, 124, 129, 133
Average crystal size; See also Stable crystal size
  addition rate and, 24, 143, 147
  in BNG model, 29–30
  in continuous nucleation, 133, 136–143, 144t,
    151t, 153–156, 161–162
  vs. critical crystal size, 3
  to critical ratio; See Average/critical size
    ratio
  crystal number and, 22, 104
  diameter of, 6
  in diffusion-controlled nucleation, 30, 37
  excess free energy and, 4–5
  growth and, 7–8
  in kinetically-controlled nucleation, 31, 33,
    94
  modeling of, 24
  Ostwald ripening and, 6
  reaction volume and, 143, 147
  residence time and, 128, 133, 143, 144t
  in seeded CSTR, 169

189